The *QuickBooks*™
Farm Accounting
CookBook™

by Mark Wilsdorf

Our Web address is:
www.goflagship.com
Visit our discussion forums & more!

Published by

FLAGSHIP TECHNOLOGIES, INC.
MADISON, MISSOURI, USA

The QuickBooks Farm Accounting Cookbook

First printing: April 1999

PUBLISHER

Flagship Technologies, Inc.
14976 Monroe Road 1039
Madison, MO 65263-9727

World Wide Web:	www.agriculture.com/markets/flagship
Email:	flagship_tech@compuserve.com
	flagship@mcmsys.com
Telephone Sales:	800-545-5380
Customer Service:	660-291-3000

COPYRIGHT

TRADEMARKS

QuickBooks, Quicken, and TurboTax are registered trademarks of Intuit. ManagePLUS is a trademark of Flagship Technologies, Inc. All other brand and product names used in this book may be trademarks, registered trademarks, or trade names of their respective holders. Flagship Technologies, Inc. is publisher and seller of the ManagePLUS software, but is not otherwise associated with any product or vendor mentioned in this book.

DISCLAIMERS

All advice and recommendations provided in this book are opinions of the author. They may not necessarily be appropriate for your specific accounting and/or tax situation. Therefore you assume all risk in interpreting, applying, and using the advice and recommendations provided herein.

Seek the advice and assistance of a professional to review your accounting software setup and operating practices, and especially for answers to tax and legal questions. No part of this book should be construed as professional advice on tax or legal matters, accounting, farm management, or any other endeavor.

Many fictitious names, including place names, are used throughout this book. All names and other likenesses to companies, persons, or places are purely coincidental and unintentional.

This project has made me fully understand why so many books are dedicated to "my wife and kids." Anyone who writes a book while still responsible for their usual day-to-day work will tell you that the extra time needed for writing has to come from somewhere. That "somewhere" is usually family time, which my wife and kids have graciously given up in the hopes that "Maybe Dad will quit being so weird when he gets the book done."

◆

This book is dedicated to my wife and kids:
Debbie, Sam, and Laura

This page is
intentionally blank.

Table of Contents

This page is
intentionally blank.

Introduction

Why do so many farmers and ranchers use QuickBooks? Because it's inexpensive? Partly. Because it's easier to learn than other programs? Probably not. I believe the main reason QuickBooks has become so popular in agriculture is that is faster and easier to use for day-to-day bookkeeping tasks: writing checks, entering deposits, paying bills, and generating basic reports of income and expenses.

QuickBooks is not an end-all, be-all farm accounting system. No one claims that it is. To make it effective for farm accounting you sometimes have to "bend" it to that purpose—use non-standard approaches to accomplish your accounting objectives. And some of the QuickBooks features which at first glance would appear to be excellent as parts of a farm accounting system don't work as expected. It's not possible to "bend" them far enough to make them work for a farm business. On the other hand, some QuickBooks features and capabilities which appear to have no use in accounting for a farming operation are actually *very* useful, once you understand how to apply them to a farm business situation.

This book was written with this landscape in mind, for anyone using Quick-Books to maintain accounting records for a farm business. The intent is to demonstrate how QuickBooks' features can be applied to a variety of farm accounting situations, and also to provide techniques you can use to "bend" Quick-Books to your farm accounting needs.

Who This Book Written "To"

This book is written to you, with the assumption that you are most likely the person directly responsible for using QuickBooks to keep the accounting records of a farm business. If that's not quite correct, I'll at least assume you have an interest in farm accounting with QuickBooks. Maybe you are an Extension agent or educator, a vocational agriculture teacher or student, or a professional accountant. (Or maybe you're an independently wealthy eccentric who prefers farm accounting to a winter vacation in Cancun or Acapulco? ...OK, so maybe not.)

The content of this book is aimed at beginning- to intermediate-level Quick-Books users, with some topics thrown in that should be of interest to advanced users.

If you are new to QuickBooks, you can quickly learn about the parts of Quick-Books you should use in a farm business (and parts you may want to avoid), how to set up a farm business company file, and how to accomplish basic tasks like entering farm income and expenses.

If you've been using QuickBooks for a while, this book should serve more as a quick reference to farm accounting techniques. Topics are presented in a "recipe" format, which means each is intended to be a self-contained block of information about a particular question or subject. Rather than reading several chapters and gathering tidbits of information here and there, you will normally be able to get your question answered by reading just a few pages.

It's a Bit Redundant
It's a Bit Redundant
It's a Bit Redundant

The downside of presenting information in a "recipe" format is that quite a bit of the information must be repeated in many places. For example, steps for adding new accounts are repeated over and over in the topics where adding accounts is necessary. This is what allows you to find all the information related to a particular topic in one section of the book. But it also makes for boring reading if you expect to read the book from cover to cover, like a novel.

You will soon learn to recognize the repeated steps (which are provided in so many places mainly to help new users) and just skim them for general information, while reading other parts of the discussion in detail.

The "Farm vs. Ranch" Thing...

I once knew a man who had eleven acres of land on the edge of a small town in the Midwest. He was a cattle buyer, and had grown up as a rancher's son somewhere on the plains—the Oklahoma panhandle I think. When he was a boy there were just two kinds of people: ranchers and farmers. And it wasn't acceptable to consider yourself some of both. You had to pick sides; you were either a rancher or a farmer (a local sentiment left over, I suppose, from the range wars of the 1880s).

Anyway, despite the fact that all the cattlemen who lived nearby considered themselves farmers, he was proud to call his vast spread (all eleven acres) a ranch. He had the ranch name painted on both doors of his pickup truck, and got a sort of distressed look if someone mentioned his "farm".

The point of this story? Regardless of what generic term I might choose for talking about the business of agricultural production I'll offend somebody—or at least leave a sour taste in someone's mouth. A few of my choices were: farm, ranch, plantation, hacienda, estate, homestead, spread, operation, and agricultural business.

I settled on "farm" (it's the easiest one to type). Wherever you see "farm" or "farm business" is used in this book, it means *any* kind of business involved in agricultural production. And if the word "farm" makes you cringe, just think ranch/plantation/hacienda/estate in its place.

A Note About Online Banking

QuickBooks supports online banking, which allows transferring funds, monitoring credit card purchases, paying bills, etc. over an Internet connection. Only a few farmers and ranchers use online banking at this time; therefore, the subject is barely mentioned throughout this book. The ideas presented here about how to handle various farm transactions apply whether you bank the "old fashioned" way or use an online banking service.

A Note About QuickBooks Versions

All menu commands and window illustrations provided in this book are based on QuickBooks 99 for Windows and QuickBooks 99 Pro for Windows, which are the most current QuickBooks versions available as of this writing. You may have to adjust commands slightly to make then work with earlier QuickBooks versions or with a Macintosh version of QuickBooks. (Specific commands for working with earlier QuickBooks versions are provided in some places.)

Comments? Suggestions? Problems?

The first edition of any book containing lots of technical material and procedural steps usually has numerous minor errors (no major ones I hope!), and this book likely will also. But besides hearing about errors, I welcome *all* comments, criticisms, and suggestions related to this book.

I may be reached at the postal or email addresses listed in the publication information at the front of the book. If you send email, include "Cookbook" in the subject line of your message and it will be routed to me. Time demands prevent me from personally responding to all email or letters, but I'll try to respond when possible. However, please don't expect a response to your personal QuickBooks-related questions or problems. (The best place to get a specific QuickBooks question answered is in one of the public Internet messaging forums described in Appendix A.)

The "Ground Rules"

Here are descriptions of elements of this book which have special meaning:

- **QuickBooks menu commands** are printed in sequence, each separated from the next by a vertical bar (|). For example, File|Open means to select the File command from QuickBooks' main menu, then the Open command from the drop-down menu that is displayed after selecting File.

- **Control key commands** are written like this: "Ctrl-R", which means to press the "R" key while holding down the key labeled Ctrl (or Control) on your keyboard.

- **Italics.** Things are *printed in italics* either to add emphasis or indicate a term that is being defined by the current sentence or paragraph.

- **Note symbol.** The symbol to the left of this paragraph flags technical notes or comments about QuickBooks operation.

- **Idea symbol.** The symbol to the left of this paragraph flags special hints or ideas about keeping records with QuickBooks.

CHAPTER 1

Basic Decisions & Things to Know

B efore you set up or begin accounting for a farm business with Quick-Books, there are some basic decisions to make about software and accounting methods.

Do I Need QuickBooks or QuickBooks Pro?

If you've decided to use QuickBooks but haven't yet purchased a copy, you need to consider whether QuickBooks or QuickBooks Pro would be the better choice. What's the difference? "About $100" would be the easy answer, but completely useless for helping you decide which program you need. The feature differences of the two QuickBooks versions are what's really important.

The first thing to understand is that QuickBooks and QuickBooks Pro are basically the same program. The general way they operate and their core accounting features are identical. Many people buy QuickBooks Pro believing the "Pro" means it's a more powerful accounting system, when really it offers just a few more features than standard QuickBooks.

As this is written QuickBooks 99 and QuickBooks Pro 99 are the current releases, and QuickBooks Pro 99 costs roughly $100 more. Most of the additional features in QuickBooks Pro 99 are of no value in a typical farm business—things like preparing job estimates, job costing, billing customers for time spent on a job, and so on—but a few of them may be useful. Here's a brief description of some of the more important ones:

◆ Export reports directly into the Microsoft Excel spreadsheet program.

You can send a copy of any QuickBooks Pro 99 report directly to a Microsoft Excel 97 or later spreadsheet. Once the report is in Excel it's possible to insert spreadsheet formulas for additional information, change fonts and colors, and rearrange the report in other ways. As you'll see in a later chapter, it also makes preparing a farm balance sheet easier. *For farm businesses this may be the most valuable additional feature of QuickBooks Pro 99.*

◆ Create letters in the Microsoft Word word processor, using your QuickBooks Pro information.

You can prepare letters for one person or an entire mailing list by identifying information in QuickBooks and having it automatically inserted in letters created in Microsoft Word 97 or later. This feature would make it easier to send a hay inventory listing to all of last year's hay customers, or a letter announcing your fall livestock sale to anyone who has purchased breeding stock from you in the past. QuickBooks Pro comes with some Microsoft Word document templates to help you get started.

◆ Multi-user capability.

Up to five people can access the same QuickBooks company file simultaneously over a network connection. If your farm business has a heavy transaction load or special information needs, and you have two or more computers linked together on a network, you may benefit from the Pro version's multi-user capability. (Using QuickBooks Pro 99 in multi-user mode requires purchasing an additional copy of QuickBooks Pro 99 for each user.)

For more details on these and the other feature differences, visit Intuit's Web site (www.intuit.com) or call Intuit at 800-446-8848. No one can predict the feature differences of future QuickBooks and QuickBooks Pro releases, but it's a sure bet that they will continue changing.

Separating Farm and Personal Funds...Or Not (How to Bend the Rules)

Standard accounting practice and the accounting equation itself (Assets = Liabilities + Owner's Equity) are based on the idea of a business entity.

A business entity is an economic unit which carries on business activity, has its own property (assets).

Proper measurement of profits and financial position requires that a business entity have its own set of accounting records—in QuickBooks terms, its own company file—and that those records be separate from records of the owner's personal finances. In other words, farm business records and personal records should not be mingled together in the same QuickBooks company file.

Separating farm and personal finances is essential if you want the records you keep to be useful for evaluating farm business profits or for preparing a balance sheet limited to farm assets and liabilities. Just as important, having separate personal financial records lets you evaluate personal/family spending habits ("Did we *really* spend that much on clothing last year?!") and aids in budgeting for family living.

But an alternate point of view is that you may not care about "standard accounting practice" or whether QuickBooks can print a balance sheet separately for the farm business. Maybe all you want is easy bookkeeping and a good set of income and expense records for tax purposes. If that's your goal, then having personal accounts (non-farm checking accounts, credit cards, etc.) in the same QuickBooks company file along with the farm business accounts is something you can certainly do. (If you have limited experience with accounting, it may even simplify the bookkeeping job.)

Whether you decide to separate farm and personal finances may depend on several factors. Here are some common ones:

* Ownership structure of the farm business.

* Whether you want your accounting system to maintain your balance sheet.

* Checkbook (checking account) management.

* Farm versus personal credit cards.

* Documentation for income tax records.

Ownership Structure of the Farm Business

If your farm business operates as a **corporation** you must keep accounting records for it that exclude personal finances, even if the corporation really just represents "you" or your family. There are employment tax, income tax, and legal reasons for doing so.

Partnership records should likewise be separate from the owners' personal finances, to have accurate income tax records and to provide all partners with information on their actual ownership interest in the farm business.

If your farm business operates as a **sole proprietorship,** you may not see any distinction between your own assets and "the farm's" assets. But there are still good management reasons why you should account for the farm business separately.

Balance Sheet Considerations

Keeping separate farm and personal records is a necessity if you expect QuickBooks to maintain correct balance sheet information for the farm business, or if you will be analyzing farm profits in relation to balance sheet changes.

But if you only use QuickBooks for income and expense records and won't be using it to prepare a balance sheet for the farm business, separating farm and personal finances is not so important. You can set up separate accounts for farm income and expenses, and for personal income and expenses, within the same QuickBooks company file. QuickBooks will still be able to prepare a tax report which selectively includes only farm business income and expense accounts. And if you do prepare a balance sheet, it will include farm and personal assets and liabilities—which may or may not be what you want, anyway.

Checkbook Management

Too often, preconceived notions about keeping separate farm and personal checking accounts become the reasons for—or against—keeping separate accounting records. But they shouldn't. Here are some things to consider:

♦ Maintaining separate farm and personal checking accounts may be worthwhile if it simplifies your record keeping, but it *is not* a requirement for keeping separate farm and personal spending and income records in QuickBooks.

It is possible to handle farm and personal spending and income records properly in QuickBooks using a single farm business checking account, without the need of a personal account. (See Keeping Non-Farm Income & Expense Records in the Farm Accounts, beginning on page .)

♦ With QuickBooks, "difficult accounting" should never be a reason for avoiding the use of two or more checking accounts. Recording transfers of funds between checking accounts is easy.

♦ Having separate farm and personal checking accounts can reduce the number of transactions you must enter in QuickBooks.

With separate checking accounts, if you only plan to keep track of farm income and expense transactions in QuickBooks you will mainly enter checks and deposits for the farm checking account...and you can mostly ignore transactions involving the personal checking account.

Having just one checking account serving both farm and personal needs would require entering *all* of the checks drawn on that account. Otherwise the checking account balance in QuickBooks would never be correct, and using QuickBooks to reconcile the account would not be possible.

Farm vs. Personal Credit Cards

Because credit cards are more often issued to individuals than to businesses, maybe you use your personal credit card for some farm-related purchases. That's fine. But to keep farm and personal finances separate in your records, you will need to learn how to enter transactions involving non-farm funds in the farm business records. See the discussion beginning on page for more information about entering farm purchases made with a personal credit card.

Documentation for Income Tax Purposes

Suppose you borrow $15,000 as a farm operating loan and the next day pay $15,000 "to boot" to trade in your old speed boat for a new one. If your records don't adequately separate farm and personal finances, an IRS auditor could make the case that the $15,000 loan was really a personal loan and disallow deducting the loan interest as a farm expense.

The best way to prevent problems like this one is to keep records which separate farm and personal spending. You should have an accurate record of owner with-drawals from the farm business and a good "paper trail" supporting transfers of funds between farm and personal use. It is *possible* to do this in a QuickBooks company file used for both farm and personal records, but the job is much easier and more automatic if the farm records are kept in their own separate company file.

Cash vs. Accrual Accounting

One of the choices you must make in any accounting system, is whether to use cash-basis or accrual accounting.

In *cash-basis accounting,* income is recorded when payment is received, and expense is recorded when payment is made. For example, repair expense is recorded when you actually write the check to pay a repair bill.

In *accrual accounting, income is recorded when it is earned and expense is recorded when it is incurred, regardless of whether any payment has been received or paid. For example, repair expense is recorded when you receive an invoice (bill) for the repairs, though weeks may pass before you actually pay the bill.*

Historically, cash-basis accounting has been the more popular method among farmers for two reasons: (1) for manual accounting in a ledger book, as most farmers were doing ten years ago, cash-basis accounting is easier, and (2) the IRS allows farmers to use cash-basis records for filing income taxes, which frequently provides significant tax benefits. But like most choices in life, the easy isn't always the most rewarding. Accrual accounting records (or cash-basis records with accrual adjustments) offer a more accurate picture of farm business profits and financial position.

Luckily, QuickBooks offers the best of both worlds. It makes the most common accrual accounting chores (accounts payable and accounts receivable transactions) so easy that almost anyone can do them. It also provides an option for getting cash-basis reports, including tax reports, whenever you want them—even from accrual accounting records.

The other alternative is to use QuickBooks strictly for cash-basis accounting, though you probably won't want to. Some of the more valuable QuickBooks features, such as managing bills, necessarily involve accrual accounting. It is possible to keep "mostly" cash-basis records with QuickBooks but also use those accrual accounting features which make the entire record keeping job easier.

How can I do accrual accounting in QuickBooks?

QuickBooks is designed to support accrual accounting but does not require it—no accounting system can! It's all up to you: *how* and *when* you enter transactions determines whether you're doing cash-basis or accrual accounting.

If you only enter checks and deposits you are doing cash-basis accounting, because transactions are only entered when money has changed hands.

But if you use QuickBooks' Bills feature to enter bills ahead of their due dates, or the Invoices feature to enter sales for which you have not yet been paid, you are doing accrual accounting. Entering a bill records the expense as of the bill's date rather than the date when you actually pay it. Entering an invoice records income as of the invoice date, though you may not receive payment for some time.

Accrual accounting involves more than accounts payable and accounts receivable (invoices and bills). However, payables and receivables make up the bulk of accrual transactions in most businesses, and QuickBooks automates their handling enough so that almost anyone can keep simple accrual records.

Single- vs. Double-entry Accounting

Single-entry accounting transactions often affect the balance of just one account—usually an income or expense account.

Double-entry accounting transactions always affect the balance of at least two accounts, which is the reason double-entry records can keep track of assets, liabilities, and changes in owner's equity in addition to incomes and expenses.

QuickBooks is a double-entry accounting system. Every transaction you enter has some effect on two or more accounts, though often QuickBooks hides this fact by making the second or "offsetting" part of the transaction behind the scenes. For example, when you enter a check QuickBooks increases the balance of the expense account(s) you selected, and also automatically creates an offset-

10

ting entry which decreases the balance of your checking account (an asset account) by the same amount.

Though QuickBooks is fundamentally a double-entry accounting system there's no problem with using it as a single-entry system. Most farmers and ranchers do. They primarily keep track of incomes and expenses, and only use the double-entry features of QuickBooks which make their accounting job easier, such as maintaining checkbook and credit card balances, and tracking loan balances and principle payments.

This page is
intentionally blank.

CHAPTER 2

Setting Up a New QuickBooks Company

This chapter contains explanations, tips, and suggestions to help you set up a farm business company file using the EasyStep Interview which QuickBooks provides for that purpose. Later chapters tell how to set up other parts of QuickBooks outside of the EasyStep interview, such as the chart of accounts.

Companies and Company Files

A **company** in QuickBooks means a single business entity, such as a farm business, a retail store, an auto repair shop, or your personal finances. All of the accounting information for a company—all transactions, the chart of accounts, preferences, and so on—is stored in a single disk file called a **company file,** which is always named with a .QBW extension, such as OURFARM.QBW.

Setting up a new company in QuickBooks creates a new company file on your hard disk. QuickBooks stores company files in the same folder (directory) where the QuickBooks program is located unless you tell it to create the file in some other location.

You can use QuickBooks to set up and maintain records for any number of companies. Each company will have a separate company (.QBW) file, which you may open in QuickBooks whenever you want to work with that company's records.

The EasyStep Interview

Beginning with version 4.0, QuickBooks has included the EasyStep Interview, a set of tabbed question-and-answer dialogs to guide you through the process of setting up a QuickBooks company.

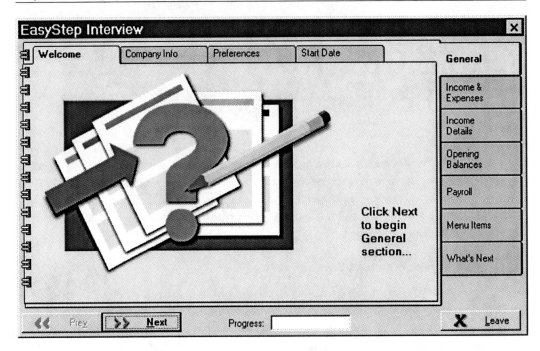

Anything you do in EasyStep could also be accomplished using QuickBooks menu commands. The big benefits of using EasyStep are that it: (1) organizes the company setup process into a series of steps in the proper order, (2) provides a good amount of explanation with each step, and (3) insulates you from the necessary QuickBooks commands by accessing them for you—something that's important when you are just getting started.

Unless you are an experienced QuickBooks user it's best to work through the entire EasyStep Interview, from beginning to end, before entering any transactions. But that doesn't mean you can't take a break! You may leave EasyStep at any time by clicking on the Leave button. If you later restart EasyStep it will return to the same page where you were working when you left.

By the way, *none* of your choices in EasyStep are "etched in stone". Any of them may be changed later, except for a few things like account names, which cannot be deleted once they've been used in transactions.

Using EasyStep is simple. Just choose an answer to the question or prompt on each page, then click the Next button to advance to the next page. To return to a previous page, click the Prev (previous) button.

To go back to a specific section of the interview, click on the large tabs along the right side of the EasyStep Interview dialog.

The following sections provide tips and information about selected EasyStep topics. They are designed so you can follow along in this book as you work through the EasyStep Interview. *If you are using a version of QuickBooks other than QuickBooks 99 some of the things discussed here may be absent from your EasyStep interview or may be arranged differently, but most of the information should apply to all QuickBooks versions.*

General Tab

> **General**

Your Company Income Tax Form

Which income tax form does your company use?

Choosing the appropriate tax form here is the first of two steps which allow QuickBooks to provide income tax reports at tax time. These reports show income and expenses grouped according to the categories on IRS tax forms, regardless of how your chart of accounts is arranged.

The second step will be when you assign an appropriate tax form line to each income and expense account. If you choose a predefined set of accounts later in the EasyStep Interview, QuickBooks will do some of this for you. For accounts you add yourself, you'll be responsible for assigning a tax line.

Select Your Type of Business

Select your type of business from the scrollable list.

Selecting a business type here allows QuickBooks to suggest a beginning set of income and expense accounts and to provide information specific to that business type.

The Farming/Ranching business type should be appropriate for most readers of this book, but not all. What's most important is to choose a business type that best matches your main business activity. Here are a couple examples:

♦ If your main activity is raising and selling vegetables and greenhouse plants at retail, then the Retail business type might be the best choice even though you may consider yourself a farmer.

♦ If much of your income comes from crop farming but most of your *bookkeeping effort* is spent on invoicing and billing customers of your custom welding and repair shop, you might choose the Service business type.

There are no clear recommendations to apply here—just choose the business type that you think best describes your business. No harm will be done by making the wrong choice, because you can add to or change any part of the company file's setup after completing EasyStep.

Your Income and Expense Accounts

Do you want to use these accounts?

After QuickBooks has created the company file EasyStep will display the predefined set of income and expense accounts for the business type you selected, and will ask if you want to use them.

◆ Choose Yes if you want to start with QuickBooks' predefined set of accounts.

◆ Choose No if you want to create your own accounts or import a set that has been exported to diskette from another QuickBooks company.

In either case, remember that you can freely add, rename, or delete accounts after completing EasyStep.

Payroll accounts: Worth keeping

If there's a chance you may someday use QuickBooks for payroll, it is best *not* to delete any of the payroll-related accounts QuickBooks creates (most have "Payroll" in the account name). These accounts are necessary for using QuickBooks' payroll features. They can be recreated manually if you happen to delete them, but not deleting them in the first place is easier!

Inventory

Does your company maintain inventory?
and
Do you want to turn on the inventory feature?

QuickBooks' inventory system was designed for retail businesses. Period. *It is very difficult to use for most farm inventories* like stored grain, feed, and growing crops or livestock. Doing so requires a lot more effort than most users are willing to spend, plus a good understanding of double-entry accounting.

But there's more to know. In some (particularly older) versions of QuickBooks you must turn on the inventory system in order to use other QuickBooks features, such as items, which you'll probably want to use. So for most farm businesses it's best to choose Yes in response to each of the questions listed above, then just bypass later EasyStep questions which prompt you to set up individual inventory items. And if part of your business activity involves buying and reselling things such as seed, feed, or even feeder livestock, then by all means choose Yes.

Tracking Segments of Your Business with "Classes"

Do you want to use classes?

Most farm business users should choose Yes here.

Classes make it possible to glean *management information* from your records in addition to financial and tax information. They are the equivalent of enterprise codes used in some other record systems, because they allow getting reports of income and expenses for the various enterprises you identify within the farm business. Also, using classes can greatly reduce the number of accounts necessary in your chart of accounts.

If you are unfamiliar with classes and want a glimpse at what they can do, skip ahead to "Setting Up Classes..." in chapter 3.

Two Ways to Handle Bills and Payments

Choose one of the two following ways to track your bills and payments.

If I could control your mouse finger for a moment I would *make* it click on the "Enter the bills first and then enter the payments later" option. This option leads to setting up one of the most valuable QuickBooks features for farm businesses: bill (accounts payable) management.

Understand that choosing "Entering the bills first..." gives you the *option* of entering bills but does not require it. You can still enter checks directly whenever you want. But being able to enter bills into QuickBooks in advance of when you pay them has several advantages. It:

♦ Calendarizes bill due dates so you always know which bills are due, and when.

♦ Lets you see cash payment requirements for the coming month or any period you choose.

♦ Lets you enter most transactions at times when you are less busy with other work and have more time for bookkeeping.

♦ Makes it easier to get bills paid on time—especially at those times when you are particularly busy—by reducing the amount of time required for preparing checks or other bill payments.

♦ Allows preparing balance sheet reports which are more accurate because they include accounts payable, the amounts you owe to others.

Bills are discussed in detail in chapter 5.

Reminders List

Think of the Reminders list as a sort of automated "notepad" which can keep you aware of accounting jobs that need to be done. Among other things, the Reminders list includes a continuously updated list of the amounts and due dates of bills to be paid, displayed in calendar order. This may be its most valuable feature for a typical farm business.

Unless you get to view the Reminders list regularly it is of little value. So choose the "At start up" option if you want QuickBooks to open the Reminders window each time you start QuickBooks.

Start Date

When selecting a start date, being overambitious is a character flaw that can come back to haunt you! If you begin using QuickBooks in August but choose a start date of January 1, you must enter *all* the transactions which have occurred since January 1 in order to have correct asset and liability balances.

Even if you plan to only keep income and expense records you probably want QuickBooks to maintain a correct balance for your farm checking account, and that's only possible if you enter all checking transactions that have occurred since the start date.

What to do? The choice is yours, but choosing a recent start date such as the beginning of the current month will require less work. If you want to have the entire year's records in QuickBooks for the purpose of preparing income tax reports, use the beginning of your tax year as the start date.

Income & Expenses Tab

There's nothing too important to do on this tab's dialogs. You can add or change income and expense accounts later, and probably more easily, using commands in the Chart of Accounts window.

> Income &
> Expenses

Income Details Tab

> Income
> Details

Receipt of Payment

Do you receive full payment at the time (or before) you provide a service or sell a product?

What this question is actually asking, is whether you want to use QuickBooks to keep track of the amounts others owe you (accounts receivable, or "customer accounts"). Your answer should at least partly depend on how well you understand the accounting that's involved.

> There are basically two reasons for keeping track of **accounts receivable**. One is so that you don't forget who owes you money, and how much. The other is so that assets are not understated when preparing a balance sheet. (Amounts others owe the farm business are considered an asset.)

Choosing "Always" prevents EasyStep from setting up the accounts receivable features, and is the appropriate choice for most farm businesses. Here are reasons why:

♦ You may want to avoid the bookkeeping effort required for maintaining customer accounts.

♦ You never send out statements of account to people or businesses who owe money to the farm business.

♦ If you mainly produce and sell commodities, it's unlikely that more than a handful of payments owed to the farm business at any time. Keeping track of them in other ways is not difficult.

♦ If you seldom prepare a balance sheet—and especially if you prepare balance sheets outside of QuickBooks—maintaining a day-to-day accounting of the amounts others owe to the farm business is less important.

If the above are reasons for *not* keeping track of accounts receivable in Quick-Books, their opposites are reasons why you may want to. If you sell to many customers on credit, need to send customer statements, or want the receivables to be automatically included in farm balance sheet prepared by QuickBooks, then choose the "Sometimes" or the "Never" option.

Items (Various Item Setup Dialogs)

Unless you've used QuickBooks before and know what items you want to set up, there's no need to enter items at this point. You may bypass all item setup in EasyStep and wait to add items later, as-needed, when you enter transactions.

Item setup is discussed in chapter 3.

Opening Balances Tab

The main purpose of this part of EasyStep is to establish beginning balances for asset and liability accounts,

Opening Balances

including customer account balances (accounts receivable, or amounts you are owed), vendor account balances (accounts payable, or amounts you owe), and other types of assets and liabilities such as fixed assets and loans.

Enter Customers

Do you have any customers who owed you money on your start date to add?

Only choose Yes here if you will be maintaining accounts receivable (customer accounts) in QuickBooks and some customers currently owe you money; that is, they have an outstanding balance on account.

Adding Vendors with Open Balances

Do you have any vendors whom you owed money on your start date to add?

Choose Yes if you want to create vendor (accounts payable) accounts and assign balances to them. Entering amounts you owe others during EasyStep is just one way to get accounts payable balances into QuickBooks as of your start date.

An alternative is to *not* enter payables here in EasyStep but to enter them later as bills, in the Enter Bills window (Activities | Enter Bills). Assign the bills a date that's on or before the company's QuickBooks start date, so the accounts payable balance will be correct as of the start date.

One of the reasons for entering beginning accounts payable balances is to allow preparing a correct balance sheet as of the start date. If you aren't concerned with being able to do this, an even simpler approach is to not enter any payables at all. Just pay the bills that were outstanding as of your start date by check as they come due, without handling them through QuickBooks' bill payment system.

Credit Card Accounts

Would you like to set up a credit card account?

You may or may not want to set up credit card accounts in EasyStep. You can always add them later, using commands in the Chart of Accounts window.

Personal credit cards in the farm business accounts?

QuickBooks warns that you should not set up accounts for *personal* credit cards in the *farm business* company file. From a proper-accounting standpoint that is true, but from a practical standpoint it makes sense to set up personal credit cards in the farm accounts in some situations—especially if you're using QuickBooks mostly for cash-basis records and aren't concerned with printing a balance sheet. See "Working with Credit Cards" in chapter 5.

Adding Lines of Credit

Do you have any lines of credit?

Your idea of a line of credit may or may not be the same as QuickBooks'. If you let EasyStep set up a line of credit account, it assigns the Credit Card account type. This lets you reconcile draws and repayments on your line of credit as if it were a checking account. This approach may be confusing if you think of a line of credit as being like other loans.

If your bank sends a monthly statement showing draws and repayments on your line of credit, then you may want to choose Yes here. will then set up the line of credit account using the Credit Card account type, and you will be able to use QuickBooks's account reconciliation features on it.

If you do not receive a monthly statement or you prefer to think of a line of credit like a standard loan, then choose No. (The next part of EasyStep will let you set up loan accounts, and you can add the line of credit account there.)

Loans and Notes Payable

Would you like to set up an account to track a loan or note payable?

This part of EasyStep lets you set up liability accounts for each outstanding farm business loan. Personal loans are not farm business liabilities, so don't set up accounts for them in the farm business company file if you plan to prepare farm balance sheets from your QuickBooks records. For more information, see the discussion of liability accounts in chapter 11.

What's the difference between a farm loan and a personal loan?

Farm business loans are those undertaken to pay farm expenses or to buy farm assets. A loan for any other purpose—to remodel your house or buy a new car, for example—would be a personal or non-farm loan.

Interest paid on non-farm loans is not deductible as a farm expense for income tax purposes.

A *litmus test:* if your tax preparer is comfortable with including a loan's interest payments as a farm expense on your tax return, the loan probably qualifies as a farm loan.

Bank Accounts

Would you like to set up a bank account?

Here, you should set up an account for the farm business checking account and any other bank accounts belonging to the farm business (money market accounts, savings accounts, etc.). Personal checking accounts should *not* be included, because they are not assets of the farm business.

Asset Accounts

Would you like to set up an asset account?

This section of EasyStep is mainly for setting up accounts for fixed assets— machinery, breeding livestock, land—and miscellaneous farm business assets such as certificates of deposit or stocks owned by the farm business (again, excluding personal asset holdings).

For now you may want to skip this section, especially if you are new to QuickBooks. You can return to this part of EasyStep later to add asset accounts, or you can add them using commands in the Chart of Accounts window.

Before you spend much time setting up asset accounts you ought to understand how QuickBooks works with assets, especially inventory-type assets like stored grain and market livestock, and how depreciation is handled. You also need to know more about your options for preparing farm balance sheets with QuickBooks. Chapter 3 provides general information about setting up asset accounts. Chapter 10 tells about techniques for handling asset accounting and depreciation.

Your Equity Accounts

To the right is a list of equity accounts which QuickBooks has added for you.

QuickBooks automatically provides a basic set of equity accounts, and lists them in this dialog window just so you'll know about them.

What does the tax form have to do with equity accounts?

QuickBooks makes an assumption about the ownership structure of your business based on the tax form you select in EasyStep, and provides a basic set of equity accounts appropriate for that form of ownership. Here are some tax form assumptions QuickBooks makes:

Form 1040 = Sole proprietorship
Form 1065 = Partnership
Form 1120 or 1120S = Corporation

Payroll Tab

Payroll setup in the EasyStep interview is relatively easy, but requires some understanding of payroll taxes, state and federal filing requirements, and so on. The QuickBooks User's Guide and Help system do a good job of describing the payroll system's setup and use, so study them before you begin.

Payroll

Payroll **is an advanced topic** which is not dealt with directly here in Volume I.

Menu Items Tab

The dialogs in this part of EasyStep ask questions about how you expect to use QuickBooks. Your answers determine how QuickBooks arranges commands in the Lists menu and the Activities menus, to get seldom-used commands out of your way.

Menu Items

What's Next Tab

This section provides instructions about other setup activities you may want to do after leaving the EasyStep Interview.

What's Next

Setting Preferences

Preferences are program settings which determine how QuickBooks operates, what information is displayed or hidden, and which accounting methods are used.

Many preferences are initially set "behind the scenes", based on you answers to questions in the EasyStep Interview. But after using QuickBooks for a while and learning more about the program, it's likely you will want to change or experiment with some of the preference settings.

This section offers tips about commonly used preference settings, and describes settings which make QuickBooks forms work in certain ways or solve problems/annoyances with how QuickBooks operates.

Forms are windows within QuickBooks where you will enter most transactions. QuickBooks forms usually mimic the appearance of real-world paper forms such as checks or invoices.

Preferences are saved separately in each QuickBooks company file. If you have more than one company (.QBW) file, the preferences may be set differently for each one.

The Preferences Window

The Preferences window is the one place in QuickBooks where all program preference settings are available—think of it as QuickBooks' main "control panel". Choose File|Preferences from QuickBooks' main menu to open the Preferences window. (Older QuickBooks versions have a Preferences menu item.)

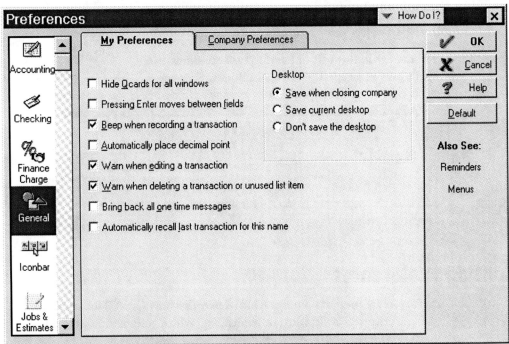

The Preferences window has two main parts: a scrollable box of icons along the left side, and a panel of preference settings on the right, occupying the larger part of the window. A different panel of preferences is displayed each time you select a different icon in the scrollable box.

The preference panels in multi-user-capable versions of QuickBooks (Pro versions 6.0 and later) are tabbed at the top, as shown in the example above. This allows separately changing some settings for individual users (the My Preferences tab) and other settings for the entire company file (the Company Preferences tab).

Important Preferences

Here are some of the important preference settings, along with where you'll find them. The preference locations are described by displaying the icon, then the preference tab's name (which may not be present in your version of QuickBooks), then the name of the preference item itself. Here's an example:

Accounting

　+ Company Preferences + Account Numbers

This translates to: locate the panel for the Account Numbers preference item by clicking on the Accounting icon in the scrollable box along the left, then on the Company Preferences tab (if one is present in your version of QuickBooks).

Require accounts in every transaction

Accounting

　+ Company Preferences + Require accounts

Turning on this preference, which is highly recommended if you want your accounting records to be complete, requires selecting an account before leaving a transaction. It assures that an account is selected in every transaction.

Allow class tracking in your transactions

Accounting

　+ Company Preferences + Use class tracking

If no Class fields appear in your QuickBooks forms, it may mean this preference has been turned off. You must turn it on if you want to use classes.

Still no Class fields on your forms?

Some QuickBooks forms do not automatically include a Class field even when class tracking has been turned on, or they may include a Class field but not where you want it. Cash Sale and Invoice forms are often set up this way—they have a Class field at the top of the form but not on each detail line. You may have to customize the form's template to show the Class field where you want it. See the QuickBooks Help system for information about customizing form templates.

Warn when you enter duplicate check numbers

Checking
+ Company Preferences + Warn about duplicate check numbers

With this preference turned on, QuickBooks will warn you if you try to enter a check that has the same check number as one already on file. This can prevent some errors, but the tradeoff is slightly slower check posting speed as Quick-Books compares the new check's number against all others. If your computer is very slow or your company file is large, turning off this preference may improve performance.

Getting rid of those Qcards...

General
+ My Preferences + Hide Qcards for all windows

Qcards are small windows of informational text that display as you move from field to field in a QuickBooks form. They can be helpful while you're learning QuickBooks but may become an annoyance later. You can turn off Qcards by selecting this preference.

Using the Enter key to move around on QuickBooks forms...

General
+ My Preferences + Pressing Enter moves between fields

Tab and Shift-Tab are the standard keys for moving between fields in Microsoft Windows. Following this standard whenever possible helps you develop habits which make it easier to learn and use a wide variety of Windows programs.

Those are the reasons for not turning on this preference...but hey, it is *your* preference! If you are more comfortable with using the Enter key to move between fields, go ahead and turn on this preference.

Automatically recalling prior transactions for easy data entry

General
+ My Preferences + Automatically recall last transaction for this name

When this preference is turned on, selecting a customer or vendor name on a form causes QuickBooks to automatically fill the form's fields with details of the most recent transaction for that customer or vendor. *This is one of QuickBooks' most valuable, useful, and highly recommended automation features!*

For instance, choosing South Bend Electric Cooperative in the Vendor field of the Write Checks window will cause QuickBooks to fill in the check with details from the most recent check to South Bend Electric Cooperative—the account, description, memo, amount, etc. All you must do to complete the new check is change any information that needs to be different, such as the dollar amount.

If the QuickBooks icon bar is not displayed...

Iconbar
+ My Preferences + Show icons and text

With this preference turned on, QuickBooks gives you one-click access to common commands from a bar of command buttons displayed just below the main menu. If your computer has a small screen you may prefer to turn off the icon bar, to make more viewable space available for forms and reports.

How to turn on the Payroll features

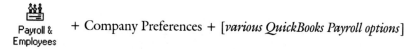
Payroll &
Employees
+ Company Preferences + [*various QuickBooks Payroll options*]

To use QuickBooks for payroll activities you must turn on one of the payroll options on this preference panel.

If you can't access the Items list...
If the Items tabs are missing from the Checks and Bills forms...
If you can't access QuickBooks' inventory features...

Purchases
& Vendors
+ Company Preferences + Inventory and purchase orders are active

Even if you don't plan to keep track of inventories in QuickBooks, this preference item *must be turned on* in some (particularly older) versions of QuickBooks if you want to use inventory-related features in other ways, such as using items to enter purchase and sale quantities in checks and bills.

If you want to automatically open the Reminders window...

Reminders
+ My Preferences + Show Reminders list when opening a company file

The Reminders window is like an automated "notepad" that QuickBooks maintains, listing current and pending bookkeeping activities such as bill due dates and amounts, important dates, and to-do notes.

To benefit the most from using the Reminders window you need to see it often. Turning on this preference item is recommended, because it causes the Reminders window to be displayed each time you open the company file.

If there's not enough (or too much) information in the Reminders window...

Reminders
+ Company Preferences + [*various Reminder settings*]

These settings control what information is displayed in the Reminders window—things like whether summary or detail information is shown, and the amount of advance notice you want before bills are due , before checks need to be printed, etc.

Below is a typical example of Reminders preferences. These will cause some reminder items to be displayed in summary form, totaled on a single line, and others such as due bills to be shown in detail, with each bill and its due date listed on a separate line. Note also that bills will appear in the Reminders window 15 days before they are due.

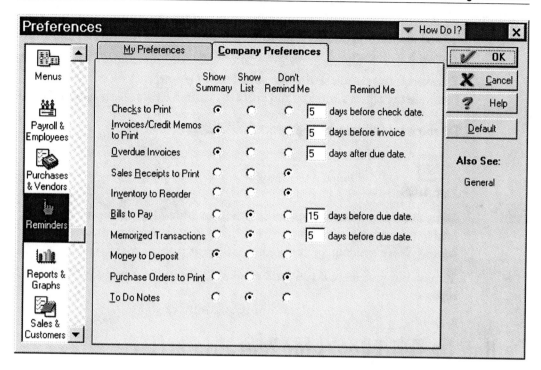

Accrual or cash-basis reports?

Reports &
Graphs
+ Company Preferences + Summary Reports Basis

Regardless of whether you keep accrual or cash-basis records, QuickBooks can prepare cash-basis reports—something you may need if you file income taxes on a cash basis, as most farmers do. This preference item sets the default method QuickBooks will use for preparing summary reports, to either Accrual or Cash. This setting does not affect transaction detail reports, 1099 reports, or sales tax liability reports.

QuickBooks has two basic report types.

Summary reports are those which only show totals and don't list individual transactions. **Detail reports** are those which list individual transactions.

♦ If you aren't certain you understand the difference between accrual and cash-basis accounting then you are probably doing cash basis accounting and should set this preference to Cash.

- ◆ If you use the accrual accounting features of QuickBooks (accounts payable and receivable) you may want to set this preference to Accrual and just over-ride it to prepare individual reports on a cash basis.

To override this default setting for an individual report, click on the Customize button in the report's window and select the desired Report Basis option.

Do you want QuickBooks to prepare 1099-MISC reports?

Tax: 1099

+ Company Preferences + Do you file 1099-MISC forms?

Many farmers and ranchers must file IRS form 1099-MISC to report amounts paid to others, such as amounts paid to individuals for custom services like hay hauling, fence building, or custom harvesting, or rent paid to landlords.

Turn on this preference if you want to allow QuickBooks to prepare 1099-MISC reports.

Setting Up QuickBooks for Your Printer(s)

Printer setup is an important part of getting QuickBooks ready to use. It in-volves choosing a printer and print options for each kind of form QuickBooks can print, such as checks, cash sales, deposits, and reports. If you are just getting started with QuickBooks, spending a little time choosing printer settings now will save time and effort each time you print a report! QuickBooks saves the printer settings, and reloads them every time the program starts up.

Printer settings apply to all companies. QuickBooks does not store different printer settings for each company.

Why Have Printer Settings?

Most Microsoft Windows programs let you specify default printer settings to control things like paper size and print orientation, but QuickBooks goes a lot farther. Here are two reasons why QuickBooks' printer setup approach is a valu-able part of the program's ease of use.

1. **It lets you store specifics about *each* type of form you print.**

   For example, QuickBooks supports printing to three different kinds of check forms. The printer setup options for checks let you select the type of check form you use, plus other specifics about how you want information printed on checks and check stubs/vouchers.

Associating specific setup information with each form type eliminates having to select those same settings each time you print a check, a cash sales receipt, a report, etc.

2. It lets you effortlessly print different forms and reports to different printers.

    You may not care much about this if you only have one printer, but if you have two or more printers it can be important.

    Some folks have a main printer which they use for most printing and a second printer dedicated to printing a particular form such as checks or invoices. They specify QuickBooks printer settings for checks so they automatically print to the second printer. Or maybe the main printer prints in black but another can print in color. In that case QuickBooks could be configured to send graphs to the color printer and all other reports to the black-only printer.

Use that old printer for printing checks?

If you have an old printer you no longer use, why not bring it out of retirement and delegate the job of printing checks to it? By leaving a supply of check forms loaded in that printer you will be ready to print checks at any time, without the bother of first having to load check forms into the printer. Setting up QuickBooks to print checks to the "XYZ printer" will automatically select that printer for printing checks each time you print them.

By the way, printing checks on an older, slower printer is a good way to keep it in service and save wear and tear on a newer printer. It's probably even a good excuse for *buying* a new printer..."Why do we need a new printer? Well Dear, so we can use the old one for printing checks..."

How to Set Up Printers in QuickBooks

1. Choose File | Printer Setup.

    QuickBooks displays the Printer Setup window.

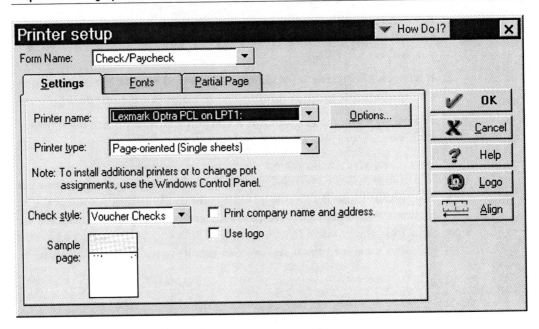

2. **Choose a form from the drop-down list in the Form Name box near the top of the window.**

   Checks/Paycheck is the selected form type in the above illustration. Fields in the lower part of the window change depending on the form type you select.

3. **Choose appropriate settings for the form you've selected.**

   Indicate the printer name, printer type, and other settings as desired. For information on these other settings, consult the QuickBooks Help system.

   Note: when you actually print a report, you can override any of the default settings you set now.

4. **Repeat steps 2 and 3 for each type of form you use in QuickBooks.**

   You can return to the Printer Setup window at any time to alter your settings.

CHAPTER 3

Setting Up Accounts, Classes, and Other QuickBooks Lists

QuickBooks keeps a lot of lists. It organizes most of the information you will use for entering transactions into lists: an accounts list, a classes list, customer and vendor lists, and many others. If you completed the EasyStep Interview, discussed in chapter 2, several of these lists may already be partly populated. EasyStep helps you create a chart of accounts and add customer and vendor names, for example.

A **chart of accounts** is a complete list of the accounting categories, or accounts, of a business—all asset, liability, equity, income, and expense accounts. In QuickBooks it is also called the **accounts list** and is displayed in the Chart of Accounts window. Throughout this book the terms *accounts list* and *chart of accounts* are used interchangeably when talking about the list of accounts in a QuickBooks company file.

You don't need to make any of these lists "complete" before entering transactions. QuickBooks lets you add to any list at any time, even while entering a transaction. Besides, adding to some lists is actually easier if you do it while entering transactions. Customers and vendors are good examples. During transaction entry is when you are most likely to have a check, invoice, or other paperwork close at hand as a source of name and address information for adding a customer or vendor to the Customers or Vendor list.

On the other hand, thoughtlessly adding entries to some lists can cause problems later: reports that don't show the information you want. The chart of accounts and classes lists are examples of lists that deserve some advance planning, so they can better serve your goals for information.

This chapter provides strategies and tips for setting up some of the more important QuickBooks lists.

About QuickBooks Lists

Why are Lists Important in QuickBooks?

Most of what's involved when you enter transactions in QuickBooks is making selections from various lists. To enter a check for example, you will typically choose an account, a vendor name (the person or business you are paying), possibly an item (to identify what you are buying or selling), and maybe a class (enterprise code). Each of these comes from a different list.

You can think of lists as categories of accounting and business information which QuickBooks maintains so that you can reuse the information by simply picking it from a list, rather than retyping it each time it is needed. Here are the lists you'll find in QuickBooks 99.

Chart of Accounts	Payment Methods
Classes	Payroll Items
Customer Messages	Reminders
Customer Types	Ship Via
Customers	Terms
Employees	Time Activities
Items	To-do Notes
Job Types	Vendor Types
Memorized Transactions	Vendors
Other Names	

List Basics

Thanks to the design of QuickBooks, nearly all features for working with lists operate the same way from list to list. Here are some of the important ones:

◆ **New list entries can be added at any time.**

You can add or edit the entries in a QuickBooks list at any time, even while entering a transaction. But as mentioned earlier, proper setup of some lists calls for a bit of advance planning. At least the basic structure of the chart of accounts, classes, and items lists should be in place before you begin entering transactions.

◆ **List entries can be renamed at any time.**

When you select a customer name in a transaction, that's what you see on the screen. But what QuickBooks actually stores in the transaction record is a *reference* to the location of the customer's name in the Customers list. This is true for all lists. If you change a customer name, account name, class name, etc., all transactions which use that name will afterward show the new name.

- ◆ **"In use" list entries cannot be deleted.**

  While even one transaction exists which uses a particular list entry, Quick-Books will not permit that entry to be deleted from the list.

- ◆ **List entries can be "hidden" by marking them inactive.**

  Beginning with QuickBooks 5.0, most lists allow marking individual list entries inactive. On QuickBooks forms where you make selections from a list, its inactive members won't be displayed. Hiding old or unused list entries is a way to reduce clutter in a list and makes it easier to use.

  Some lists allow merging (combining) list entries.

  If you decide that two account names, customer names, or vendor names represent the same information, QuickBooks will let you merge them—join them under a single name. When you do, QuickBooks automatically makes the necessary changes so that all transactions refer to the new name.

- ◆ **The best place to do list management chores is in the list's window.**

  Every list has its own window, and most list windows are accessible from the Lists item in the main menu. For example, choosing Lists|Chart of Accounts will open the Chart of Accounts window.

  A list window displays all of the list's entries and provides list management commands for chores like adding or editing list entries, or printing reports related to the list. The commands may be accessed from command buttons along the lower edge of the list window or from the Edit item in the main menu.

The contents of the Edit menu changes each time you activate a different window in QuickBooks, so editing commands specific to the currently-active window are always available there.

Setting Up a Chart of Accounts

Account Basics

Traditional accounting defines five basic types of accounts, all of which are used in QuickBooks: Asset, Liability, Equity, Income, and Expense.

Income and Expense Accounts

Almost everyone is familiar with *income* accounts and *expense* accounts. They are categories for tracking the amount of income and expense experienced by a business over a period of time.

Income and expense accounts are often called *profit and loss accounts* or *P&L accounts* because they appear on profit and loss reports (also known as an income statement). A profit and loss report lists income and expenses for a particular period of time and calculates net income by subtracting total expense from total income.

If you use QuickBooks primarily as a single-entry accounting system, you will mostly work with income and expense accounts. (A few asset, liability, and equity accounts are necessary for operating QuickBooks however, such as an asset account to represent the farm checking account.)

Asset, Liability, and Equity Accounts

Together, asset, liability, and equity accounts are often referred to as *balance sheet* accounts. A balance sheet, sometimes called a financial statement, consists of the current balances of these three accounts related in this way:

$$\text{Assets - Liabilities = Equity}$$

This is one form of what accountants call the *accounting equation*.

Assets represent the things of value that the business possesses—land, buildings, machinery, livestock, supplies, cash, certificates of deposit, etc.

Liabilities represent financial obligations (debts) to others who are not owners or partners in the business, such as loans, notes, lines of credit, and credit cards. You might say that liabilities represent the portion of the business that is owned by or owed to others.

Equity represents the value of the owner's stake in the business. As the accounting equation above shows, equity is a residual or "left over" amount. It's the amount that's left over after subtracting liabilities from the value of the business' assets.

Because QuickBooks is designed for double-entry accounting it maintains asset, liability, and equity account balances as well as incomes and expenses. Every transaction entry which has some effect on a farm business asset or liability will change the balance of an asset or liability account.

Guidelines for Creating Accounts in QuickBooks

Here are some of the basic concepts that are important for organizing a chart of accounts in QuickBooks.

1. **Let each account represent a broad category of information.**

   Avoid having a large number of special-purpose accounts. The goal should be to keep your chart of accounts as simple as possible. This makes it easier to locate accounts during transaction entry, helps you remember the purpose of each account, and keeps your reports simple.

   For example, one Repairs account may be enough. You shouldn't need a dozen different ones with names like Tractor Repairs, Barn Repairs, Pickup Repairs, and Combine Repairs.

   If you want that much detail on repair expenses for *specific* items, use classes instead of adding more accounts. Using classes to represent enterprises, specific machines, or other special items will help to keep your chart of accounts simple. If you only want more information on *general* categories of repair expense, set up some subaccounts of Repairs such as Equipment Repairs and Building Repairs.

   Classes are similar to accounts, but separate. Think of them as a second set of accounts that you may apply to transactions in addition to the regular accounts. They allow grouping and summarizing income and expenses in ways that are *independent* of the chart of accounts, usually related to specific enterprises within the farm business. See page 59 for more information about setting up and using classes.

2. *Resist the temptation* to set up subaccounts for farm enterprises.

   In almost all cases you should identify enterprises using classes, not accounts. This keeps your chart of accounts simple and brief. It also makes reports less complicated and easier to prepare, especially when you want a report for a specific enterprise.

   The basic idea is summed up as...

The Golden Rule of Account and Class Setup
Use accounts to identify the "what" of a transaction and classes to identify the "why" or "what for".

Here's an example of this Golden Rule: if you buy fertilizer to apply to corn, the "what" is fertilizer; the "what for" is corn. So fertilizer should be an account and corn should be a class.

Creating lots of enterprise-specific subaccounts is a common mistake among new QuickBooks users. To illustrate the problem, consider this fragment from a simple chart of accounts:

> **Fertilizer**
> **Chemicals**
> **Storage & Warehousing**

Now here's part of the same chart of accounts, but with subaccounts added to represent Corn and Soybean enterprises:

> **Fertilizer**
> > **Corn Fertilizer**
> > **Soybean Fertilizer**
> **Chemicals**
> > **Corn Chemicals**
> > **Soybean Chemicals**
> **Storage & Warehousing**
> > **Corn Storage Fees**
> > **Soybean Storage Fees**

The account list quickly grows large and unwieldy when enterprise-specific subaccounts are added. It's also redundant and time consuming to set up, because Corn- and Soybean-specific subaccounts must be added in many places.

But consider about how things would be different if the Class list contained the following two classes.

> **Corn**
> **Soybeans**

If these were used in transactions instead of the enterprise-specific subaccounts, the first (simple) chart of accounts would do. The result would be just as much information but with much less clutter and confusion.

If your chart of accounts has a lot of enterprise-specific subaccounts and you'd like to switch to using classes instead, see "Managing QuickBooks Lists" in chapter 9. It tells how to merge subaccounts into their parent account, while preserving the enterprise information by adding classes.

Don't think this Golden Rule is ironclad. Like most rules it's meant to be broken in some situations, usually when you want to control how information is arranged in reports. Here is an example, part of the income section from a chart of accounts, with Corn Sales and Soybean Sales subaccounts.

> **Sales**
>> **Grain**
>>> **Corn Sales**
>>> **Soybean Sales**

If Corn and Soybeans *classes* are available, what's the purpose of having Corn Sales and Soybean Sales subaccounts? The simple reason for using enterprise-specific subaccounts here is that it will cause Corn Sales and Soybean Sales totals to appear directly on profit and loss reports. If only classes were used, a separate report would be necessary to get totals for corn sales and soybean sales.

Another situation where enterprise-specific subaccounts are appropriate is for showing details about farm inventories on balance sheet reports. For example, a Hay Inventory asset account might have Alfalfa, Clover, Grass/Legume, and Prairie Hay subaccounts.

3. **Let accounts represent categories of income and expense that are meaningful to *you* rather than to the IRS.**

Income tax preparation should never be the main factor controlling how your chart of accounts is organized. QuickBooks frees you from any need to have your accounts mimic the categories on IRS tax forms. You can get the tax reports you need—with income and expenses grouped in the appropriate tax form categories—regardless of how your chart of accounts is arranged or the account names you have used. For more information see Preparing Income Tax Reports, in chapter 12.

4. **Before adding a new account, give it some thought.**

QuickBooks lets you set up new accounts at any time, even in the middle of entering a transaction, but don't be hasty about adding accounts. You should always be able answer these questions when adding a new account:

- **Do I really need a new account or should I use an existing account?**

    As mentioned earlier, your chart of accounts will be easier to use if you keep it simple. Avoid creating a new account for every purpose.

- **Should I add the new account as a main account or as a subaccount?**

    If the account you want to add relates to an existing account, consider adding it as a subaccount. If you already have an account called Utilities, with subaccounts named Electricity and Telephone, you might add Rural Water as a subaccount of Utilities.

- How will the new account affect the way information is shown in reports?

    Using subaccounts keeps related information grouped together, both in your chart of accounts and on reports, and can make reports easier to read and understand.

QuickBooks Account Types and Examples

Every account you set up in QuickBooks must be assigned an account type. This section gives you an overview of QuickBooks account types, describes how they apply to farm and ranch situations, and provides some typical farm business examples of each type. The goal is not to provide a "cookie-cutter" chart of accounts but to give you ideas for building your own.

If you are new to QuickBooks some of the examples and explanations in this section may not make a lot of sense to you just yet. You may want to just skim this section for general information, and come back to it later after you've learned more about QuickBooks.

Current Asset Account Types

Current assets are "cash or near cash" assets, including cash, bank accounts, marketable securities, accounts receivable, prepaid expenses, and inventories of grain, market livestock, and supplies.

Traditionally, current assets includes things you expect to sell or use up in less than a year. Another important criteria is that you can convert current assets into cash (sell them off) without reducing the farm's ability to produce output. For example, market livestock are a current asset: selling market hogs does not cut into the farm's ability to produce pork in the future. But selling part of the sow herd reduces the farm's pork production capacity. (Sows or any other breeding livestock are fixed assets, not current assets.)

Type	Examples
Bank	Checking, money market accounts, savings accounts

Having at least one bank account is a basic requirement of using QuickBooks. If you don't yet have a bank account set up, the first time you try to enter a check or deposit QuickBooks will prompt you to create one.

This account type can be assigned to any account you have at a financial institution, and also to petty cash accounts.

Type	Examples
Accounts Receivable	Accounts Receivable

QuickBooks is designed to work with *just one* account of the Accounts Receivable type, and automatically sets up the account the first time you try to create an invoice. *Do not create additional Accounts Receivable type accounts unless you are an accounting expert and you feel it is absolutely necessary!* (In other words, don't do it!)

Type	Examples
Other Current Asset	Prepaid expenses; inventories of grain, market livestock, feed, supplies, etc.

This is a "catch all" current asset type. Assign it to any current assets which are neither Accounts Receivable nor Bank accounts. In a farm business Other Current Assets typically are inventories of grain, market livestock, and other farm production.

It is impossible to provide examples of Other Current Asset accounts which suit every situation because the nature of farm inventories is so different from one farm to the next. Also, the accounts you need will depend on the kind of balance sheet reports you want to produce—market value or book value—and how you choose to keep track of farm inventories in QuickBooks, if at all! (If you won't be producing balance sheets or keeping track of farm inventories in QuickBooks you will need few, if any, farm inventory accounts.)

Here are some detailed examples of Other Current Asset accounts to jump start your thinking about the ones you may need:

> Stored Grain Inventory
> > Corn
> > Soybeans
> > Wheat
> Value of Growing Crops
> > Corn
> > Soybeans
> > Wheat
> Market Livestock Inventory
> > Hogs

> Birth to Weaning
> Weaning to 150 lbs.
> 150 lbs. to Market
> Dairy Cattle
>> Calves
>> Heifers
> Beef Cattle
>> Calves, Pre-weaning
>> Stocker Calves
>> Feedlot Cattle

Feed Inventory
> Grain for Feed
> Complete Feeds (purchased)
> Mineral Supplement
> Protein Supplement
> Other

Hay Inventory
> Alfalfa
>> 1st cutting
>> 2nd cutting
>> 3rd cutting
>> 4th cutting
> Clover
> Grass
> Grass/Legume

Supplies Inventory
> Chemicals
> Fuel
>> Gasoline
>> Diesel
> Seed
> Supplies

Fixed Asset Account Types

Accountants often refer to fixed assets as "plant and equipment". They are the long-run tools of production owned and used by the farm business for more than one year or production cycle. Traditionally, assets you expect to hold for more than one year are considered fixed assets. Examples are machinery, breeding livestock, and land.

Type	Examples
Fixed Asset	Machinery, breeding livestock, land

Maintaining fixed asset accounts is only necessary if you want to print balance sheets directly from QuickBooks. Some people prefer to prepare balance sheets another way—with a spreadsheet program for instance—and don't track fixed assets in QuickBooks. If that's your preference, you only need a few accounts for recording fixed asset purchases and sales, for tax purposes:

Fixed Asset Changes
Purchases
Sales/Disposals

At tax time you can print a report based on these accounts, that will have most of the information your tax preparer needs about purchases, sales, and disposals of fixed assets.

If you will be preparing balance sheet reports with QuickBooks, before setting up fixed asset accounts you should decide whether you want assets to be listed at book value or market value on the balance sheets. The farm's fixed asset accounts may need to be structured a bit differently depending on which you choose.

For *market value balance sheets* you can use one account to represent each fixed asset, or each group of fixed assets if you want them grouped on the balance sheet. Your fixed asset accounts should have the same amount of detail as you want shown on balance sheet reports. Actually, fewer accounts is better, because each time you prepare a balance sheet you'll have to manually update each one to reflect current market values.

Here is a simple set of fixed asset accounts, each representing a broad group of fixed assets:

Breeding Livestock
Machinery
Land & Improvements

And here's how subaccounts might be used to provide more detail on balance sheet reports:

Breeding Livestock
    Beef Cows
    Other
Machinery
    JD 7700 1997
    Kinze Planter 1996
    Case-IH Disk 1994

> Buildings
>> Machine Shed 1995
>> Grain Leg & Pit 1994
> Land
>> Home farm, 220 acres
>> Smith place, 120 acres

See chapter 10, Farm Asset Basics, for more information about preparing market value balance sheets.

QuickBooks account names can be up to 31 characters long. Use that space to make your account names more descriptive so they better identify the assets (or whatever) they represent. Things like the date or year of purchase, model names or numbers, or the number of animals in a purchased group can all be parts of the account name.

If you want to prepare *book value balance sheets,* either a simple set of fixed asset accounts (with most accounts representing groups of assets) or a more detailed set will do. In either case you should add subaccounts for keeping track of the assets' original cost (basis) and accumulated depreciation. Here's a set of accounts like the detailed set shown earlier, but with Basis and Accum.Depr. (accumulated depreciation) subaccounts added...except for Land, because land is not depreciated.

> Breeding Livestock
>> 1998 Angus Cows 11
>>> Basis
>>> . Depr.
>> 1999 Gelb/AngXCows 23
>>> Basis
>>> Accum. Depr.
> Machinery
>> JD 7700 1997
>>> Basis
>>> Accum. Depr.
>> Kinze Planter 1996
>>> Basis
>>> Accum. Depr.
>> Case-IH Disk 1994
>>> Basis
>>> Accum. Depr.
> Buildings
>> Machine Shed 1995
>>> Basis

Accum. Depr.
Grain Leg & Pit 1994
Basis
Accum. Depr.
Land
Home farm, 220 acres
Smith place, 120 acres

Tracking Fixed Assets...by Payee Name?

The downside of having so many detail accounts is that they'll all appear on balance sheet reports prepared by QuickBooks! That may be more detail than you want the balance sheet to show.

One alternative, depending on your version of QuickBooks, may be to set up a smaller number of accounts and let each account represent a group of assets—such as the entire cow herd, or all of your machinery. You can track individual assets within these accounts by storing the asset's name in the Payee Name field of the fixed asset account Register. Then, to see all transactions for a particular asset you can get a , filtered for that asset's Payee Name. Consult the QuickBooks Help system for details on this method.

Type	Examples
Other Asset	Land, long-term notes receivable

Other Asset is a "catch all" account type for fixed assets. Some QuickBooks users apply this type to all non-depreciable fixed assets such as land, and reserve the Fixed Asset account type for depreciable fixed assets. The choice is yours.

Intermediate vs. Long-term Fixed Assets

Some people like to separate intermediate and long-term fixed assets on balance sheet reports, but QuickBooks does not automatically make that distinction. If you want intermediate and long-term assets to be grouped separately on balance sheets, set up separate parent accounts for them and subaccounts for individual assets or asset groups, like this:

Intermediate Assets
Breeding livestock
Machinery
Long-Term Assets

Buildings
Land

For more asset detail you can add one or more additional levels of sub-accounts.

Current Liability Account Types

Current liabilities are debts which must be paid within a year. Typical examples are accounts payable (bills for feed, veterinary services, repairs, electricity, fuel, etc.), credit cards, and operating notes or other short-term lines of credit.

Type	Examples
Accounts Payable	Accounts Payable

QuickBooks is designed to work with *just one* account of the Accounts Payable type, and automatically sets up the account the first time you try to enter a bill. *Do not create additional Accounts Payable type accounts unless you are an accounting expert and you feel it is absolutely necessary!* (In other words, don't do it.)

Type	Examples
Sales Tax Payable	Sales Tax Payable

QuickBooks automatically sets up a Sales Tax Payable account if you've selected the sales tax collection option in the interview or the Preferences window.

Note: Sales Tax Payable accounts can only be created by QuickBooks, because this account type is excluded from the account types list.

Type	Examples
Credit Card	Credit card accounts

Set up a separate Credit Card type account for each farm business credit card. Personal credit cards should only be included in the farm business records if you have a clear understanding of how they relate to the farm business. (See Working with Credit Cards in chapter 5.) Examples:

FirstBank VISA
BankOnUs Mastercard

Type	Examples
Other Current Liability	Short-term notes payable, lines of credit, master notes.

This is a "catch all" current liability type. Use it when setting up current liability accounts which are neither Accounts Payable nor Credit Card types. For easier control and reporting related to loans, always set up a separate liability account for each loan—never lump them together in a single account. Examples:

Credit Line $32,000
Operating Note $10,000
Hay Rake Note, 6-Month $5,000

Naming loan accounts...

Including the original amount borrowed or the credit line maximum in the account name is a handy way to identify loan accounts.

Long-Term Liability Account Types

Long-term liabilities are debts to be paid off over a period of more than one year. Any loan with a repayment period of more than one year is a long-term liability. Typical examples are loans for machinery, breeding livestock, and land.

Type	Examples
Long Term Liability	Land loans, machinery loans, etc.

For easier control and reporting related to loans, always set up a separate liability account for each loan—never lump them together in a single account. Examples:

Tractor Note-AgriMech $55,000
Land Note-1st America $120,000

Some people like to separate **intermediate liabilities** and **long-term liabilities** on balance sheet reports, but QuickBooks does not automatically make that distinction. If you want intermediate and long-term liabilities to be grouped separately on balance sheets, set up separate parent accounts for both and enter the individual liabilities (loans) as subaccounts of the two parent accounts.

47

Intermediate Liabilities
Tractor Note-AgriMech $55,000
Long-Term Liabilities
Land Note-1st America $120,000

Equity Account Types

Equity is the difference between the value of a business' assets and liabilities. It represents the value of the owners' or stockholders' investment in the business.

Type	Examples
Equity	Equity, capital, and drawing accounts

As you will see below, QuickBooks may create several different Equity type accounts when you set up a new QuickBooks company. At least two equity accounts, Opening Balance Equity and Retained Earnings, are always created. Other equity accounts are created during the Interview based on the income tax form you select.

QuickBooks uses the **Opening Balance Equity** account (actually named Opening Bal Equity) as the offsetting account when you assign an opening balance to asset and liability accounts. But this mostly happens when you are first setting up a company file. After that you may use the account any way you like. Opening Balance Equity is often used as a "catch-all" equity account for account balance adjustments—when you need to change an account balance but "throw away" the offsetting part of the entry by posting it to an equity account.

Retained Earnings is a special account. It does not have a Register (a window which shows the account's transaction history), nor does the Chart of Accounts window display a balance for Retained Earnings as it does for other equity accounts. The intended purpose of Retained Earnings is to represent the accumulated value of net income from all prior years of operating the business, on the balance sheet. QuickBooks essentially uses it to adjust the balance sheet's equity total to equal the difference between assets and liabilities (to keep the balance sheet "in balance" by adhering to the basic accounting equation: Assets - Liabilities = Equity).

QuickBooks will automatically create equity accounts *only* when you first set up a new company, during the EasyStep Interview. (Selecting a different tax form later will not cause any equity accounts to be added or changed.)

If the tax form you select during EasyStep is Form 1040 (for sole proprietorships), these equity accounts will be created:

> Opening Bal Equity
> Owner's Capital
> > Draws
> > Investments
> Retained Earnings

The Draws and Investments subaccounts allow keeping track of funds that are withdrawn from or added to the farm business.

Renaming Equity Accounts

You may rename any QuickBooks account at any time. So feel free to rename accounts using names that are easier to remember or that better describe the account's purpose. For example, you might rename the Owner's Capital accounts shown above like this:

> Capital, John Doe
> > Draws
> > Additions

If the tax form you select during EasyStep is Form 1065 (for partnerships), QuickBooks will create these equity accounts:

> Opening Bal Equity
> Partner One Equity
> > Partner One Draws
> > Partner One Earning
> > Partner One Investments
> Partner Two Equity
> > Partner Two Draws
> > Partner Two Earning
> > Partner Two Investments
> Retained Earnings

The Draws and Investments subaccounts allow keeping track of funds each partner has withdrawn from or added to the business. The purpose of the Earning subaccounts is to allow assigning a portion of net income to each partner, from Retained Earnings.

If the tax form you select during EasyStep is Form 1120 (for corporations) or 1120S (for Subchapter S corporations), QuickBooks will create these equity accounts:

Capital Stock
Opening Bal Equity
Retained Earnings

You may use QuickBooks as a stock ledger if the farm corporation has only a few shareholders. Add subaccounts to Capital Stock to keep track of the shares outstanding to each shareholder:

Capital Stock
    Stock, John Doe
    Stock, Mary Doe
    Stock, Bill Johnson

Income Account Types

Income accounts are for recording income from sales of farm production.

Type	Examples
Income	Grain, market livestock, or produce sales; government program payments; patronage dividends

One of the choices available during the Interview is to let QuickBooks supply a preset farming/ranching chart of accounts. Though it can't suit everyone, the preset chart of accounts is a good starting point for building your own chart of accounts. It contains many of the income accounts you will need, plus you can add others or delete any you don't need.

How you structure the income section of your chart of accounts determines the amount of income detail that will be on profit and loss reports. Using enterprise-specific subaccounts is generally a bad idea in most parts of the chart of accounts. But having a few in the income section is OK, to get a more detailed breakdown of income on profit and loss reports. Here are a variety of income account examples, including several that use enterprise-specific subaccounts to categorize income in greater detail. (To take a less detailed approach with your own chart of accounts, just use fewer subaccount levels.)

Custom Work Income
  Tillage
  Planting
  Baling
  Combining
  Trucking
Sales
  Grain
    Corn
    Soybeans
    Wheat
  Hay
    Alfalfa
    Grass
  Cattle
    Weaned Calves
    Stockers
      Raised
      Purchased for Resale
        Sales
        Cost
    Fed Cattle
      Raised
      Purchased for Resale
        Sales
        Cost
  Dairy
    Calves
    Breeding Heifers, Raised
    Milk
  Hogs
    Feeder Pigs
    Market Hogs

Why so many subaccounts?

If you are a new QuickBooks user you may worry that multiple subaccount levels will make transaction entry more difficult. But it doesn't, once you are used to it. And using subaccounts is worthwhile because it yields additional information—more detailed categories of income and expense.

Type	Examples
Other Income	Unusual or extraordinary income

This account type is for recording income that is unusual or extraordinary income—income that's outside of the normal flow of income from farm operations. An example of Other Income would be payments received from a levee maintenance district for soil from one of your fields, bought to increase the height of a river levee.

The real reason for having an Other Income account comes into play when you create a profit and loss report. Because Income and Other Income are totaled separately, the profit and loss report can show both ordinary net income and net income. *Ordinary net income* is the amount of net income the business would have earned from normal operations, without the benefit of the extraordinary additional income recorded as Other Income. *Net income* includes all income.

Expense Account Types

Expense accounts are for recording farm business expenses.

Type	Examples
Expense	Feed, seed, fertilizer, supplies, rent

QuickBooks' preset farming/ranching chart of accounts, one of the choices available during the EasyStep interview, is a good starting point for building your own farm business chart of accounts. It contains many of the expense accounts you will need, plus you can add others or delete any you don't need.

Accounts you may want to keep...

If there's a chance you may someday want to use QuickBooks' payroll features, it's best to keep any of the payroll accounts supplied in Quick-Books' preset chart of accounts (most have "Payroll" in the account name). Those accounts are necessary for using the payroll features and, though they can be recreated manually, it's much easier not to delete them in the first place!

One thing the farming/ranching preset chart of accounts does not illustrate well, is how you might use subaccounts to break down expense information in more detail. As you consider the possibilities though, remember this: the only good reason for adding more detail accounts is if it will result in useful management information. It's best to start with a chart of accounts that seems "too simple" and add subaccounts later when you have a fuller understanding of how Quick-Books works and of your own information needs.

Here are some examples showing how you might use subaccounts in various ways for more detailed expense information.

Chemicals Expense
  Herbicide
  Insecticide
  Forage Preservative/Innoculant
  Other
Feed Expense
  Grain
  Hay
    Alfalfa
    Legume
    Grass
  Complete Feeds
    Calf Creep
    Receiving Ration
  Premix
  Mineral
  Supplements
    Protein Mix 40%
    SBOM 48%
    Soybean Hulls
    Corn Gluten Feed
    Other
Fuel & Lubricants
  Gasoline
  Diesel
  LP Gas
  Lube Oil
  Grease
Interest Expense
  Interest & Finance Charges
  Mortgage Interest
Machine & Labor Hired
  General Labor
  Hay Baling Hired
  Hay Hauling
  Spraying

Other Custom Application
Veterinary Expense
Vet. Fees
Vet. Supplies & Medicine
Implants/Growth Promotants

Type	Examples
Costs of Goods Sold	Costs of Goods Sold

This is a special expense (actually, contra-income) account type that's meant for use with QuickBooks' inventory features. QuickBooks normally creates a Cost of Goods Sold account if you enter a sale of Inventory Part type items. When Inventory Part items are sold, QuickBooks will normally post their purchase cost to the Cost of Goods Sold account.

Type	Examples
Other Expense	Unusual or extraordinary farm expenses

This account type is for recording expenses which are unusual or extraordinary—outside of the normal flow of farm business expenses. An example of Other Expense would be the cost of major building repairs due to uninsured damage from an earthquake.

The real reason for having an Other Expense account comes into play when you create a profit and loss report. Because Expense and Other Expense are totaled separately, the profit and loss report can show both ordinary net income and net income. *Ordinary net income* is the amount of net income the business would have earned from normal operations, without the unusual or extraordinary expenses recorded as Other Expense. *Net income* includes all income and expense.

Getting Started with Your Own Chart of Accounts

The organization of your chart of accounts determines (1) how easy it is to use, (2) how your reports will be structured, and (3) how effectively income and expenses will be categorized in ways that yield useful management information. These facts make the chart of accounts the most important list you will work with in QuickBooks and are good reasons to spend at least a little time planning how it is organized.

Here are some tips for planning and setting up your chart of accounts.

♦ **Set up the main structure of your chart of accounts—the main accounts and subaccounts—before you start entering transactions.**

True, if you are new to QuickBooks this may not be easy. You won't be familiar enough with how QuickBooks works to make informed decisions about the kinds of accounts you need and how to arrange them.

If you are setting up a new QuickBooks company file, a good way to start is by selecting the preset farming/ranching chart of accounts in the Interview. With those accounts as a base to work from, you can add other accounts you think you'll need and delete accounts you don't expect to use.

A good time to review your chart of accounts and make changes is before you begin entering transactions for each new accounting year. If you need to reorganize, add, or hide some accounts, doing it then will let you to keep the structure of your accounts list mostly the same for all transactions entered in the new year.

♦ **You *do not* need to set up every possible account you may need, before entering transactions.**

This idea may seem counter to the preceding one, but it's not. The point is that you can build your chart of accounts over a period of time by adding new accounts whenever you see the need for them.

If you plan to use QuickBooks for *single-entry accounting* records (mostly income and expenses), you obviously need to think about how to organize income and expense accounts. Setting up accounts for the main income and expense categories should be enough, before you begin entering transactions. You can add subaccounts and "flesh out" the chart of accounts later as the need arises.

Because single-entry records don't directly support preparing a balance sheet, you'll only need to set up the asset, liability, and equity accounts necessary for running QuickBooks. The most important of these accounts are normally established during the EasyStep Interview, such as an asset account to represent the farm checking account. But you may also want to set up others as a matter of convenience, such as individual liability accounts for each farm loan.

The other approach is to use QuickBooks for *double-entry accounting* records, which track asset and liability changes in addition to income and expenses, and therefore support preparing balance sheet reports. In addition to income and expense accounts you need to set up accounts for the farm's assets and liabilities, with opening balances which are correct as of your QuickBooks start date. This *does not* mean, however, that every one of these accounts must be in place before you start entering transactions.

If you are in a hurry to get started, just set up the necessary accounts now, such as the farm checking account, and put off setting up others until you need them for a transaction entry or want to prepare a balance sheet. (Most accounts representing fixed assets like machinery, breeding livestock, and land, won't be necessary until you want to prepare a balance sheet report.)

♦ **Don't lose sleep over deciding how to set up accounts.**

You're never entirely "stuck" with a particular account setup. QuickBooks lets you move accounts to different locations within the chart of accounts, hide accounts you no longer use, and merge (combine) accounts under a single account name. You can even make a subaccount into a main account, or vice versa.

♦ **Don't spend too much time trying to set up the "perfect" chart of accounts.**

There is no such thing! Your chart of accounts will inevitably change as your business and information needs change and as you gain experience with QuickBooks. *Expect to redo part of your account setup each year for the first year or so of using QuickBooks.* Just consider it part of your "tuition cost" for learning QuickBooks!

How to Set Up New Accounts

Though you may add accounts from many places in QuickBooks—from any form where you can select an account—the Chart of Accounts window is the best place to start if you want to add new accounts. Besides showing the entire accounts list it gives you access to all of the commands for managing the list, including adding, editing, and deleting accounts. Here's how to add a new account:

1. **Open the Chart of Accounts window.**

   Choosing Lists | Chart of Accounts is one way to do this.

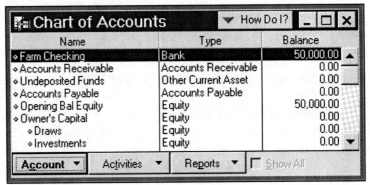

2. **Click on the Account button at the bottom of the window.**

A pop-up menu of commands for working with accounts will be displayed.

3. **Click on New in the pop-up menu.**

QuickBooks will open the New Account window. (This same window will open if you choose to add a new account while entering a transaction.)

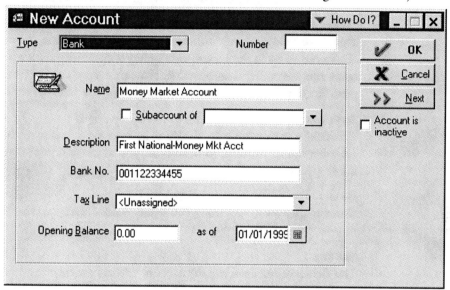

4. **Fill in information about the new account.**

Here are descriptions of the New Account window's fields.

Type

The account type you select determines whether the account is an asset, liability, equity, income, or expense account, and its normal order of appearance in the Chart of Accounts window. The reason there are more than five account types to choose from is that QuickBooks handles accounts of each type in slightly different ways. Account types were discussed earlier in this chapter.

Number

Account numbers are completely unnecessary in QuickBooks. However, if you are accustomed to a particular account numbering scheme you can use it by entering a number for each account here in the Number field. *If the Number field is absent and you want to use account numbers, you must turn on the account numbers feature by choosing the "Use account numbers" setting in the File | Preferences | Accounting dialog .*

Name

This is the name that will appear in the Chart of Accounts window and also the one you will type or select when entering a transaction. Account names can be up to 31 characters long, so don't scrimp! Make them descriptive enough that you can easily remember each account's purpose.

Subaccount of

To make your new account a subaccount of another, click on the check box and select the name of the parent account in the box to the right.

Description

The Name field is usually sufficient to identify an account, so you usually don't need to fill in the Description unless you identify accounts by an account number rather than by name. Note that there's a File|Preferences|Reports & Graphs option which allows including the Description on reports.

Bank No.

This field is present only for the Bank account type. It is required only if you are using online banking, in which case you must enter your bank account number here.

Note

You may use this for anything you want. It does not appear on any financial reports.

Tax line

This field lets you associate the account with a particular income tax form line for the purpose of generating income tax reports. You won't assign a tax line to all accounts, mainly just income and expense accounts. And you don't have to assign tax lines when first setting up accounts, so long as the job gets done before preparing tax reports.

Opening balance and as of

The Opening balance field lets you supply an opening balance when setting up asset, liability, or equity accounts. (It is not available when setting up other account types.) This should be the account's balance as of your QuickBooks start date or the date you enter in the **as of** field.

Except when first setting up a company in QuickBooks, leave the opening balance of new accounts at zero. When you enter an opening balance an offsetting amount gets posted to the Opening Balance Equity account. For assets or liabilities acquired after your QuickBooks start date this is *not* what should happen. Instead, new asset and liability accounts should be established with an opening balance of zero. The beginning balance will normally be transferred in automatically by the first transaction which references the account, such as a check written to purchase a new piece of machinery.

Account is inactive

Check this box to mark the account inactive, which means it won't be shown in account selection boxes on QuickBooks forms. Hiding accounts you no longer use reduces clutter in the accounts list and makes it easier to use. You may reactivate an account at any time by deselecting this check box. (To view inactive accounts, click on the Show All option in the Chart of Accounts window.)

5. **Click OK or Next to save the new account in the accounts list.**

   Clicking OK saves the account and closes the New Account window. Clicking Next keeps the New Account window open for entering another account.

Setting Up Classes: Enterprise Information and More

The QuickBooks Classes list is like a second set of accounts you can apply to transactions *in addition to* the regular accounts. Using classes lets you get reports and analysis of the farm business records according to categories that are completely independent of the main chart of accounts.

Classes are most often used as enterprise codes, to match transactions with specific farm enterprises and to get income and expense reports focused on those enterprises rather than the farm business as a whole. Another benefit of using classes is that it lets you gather enterprise information without overloading the chart of accounts with enterprise-specific subaccounts. That keeps the chart of accounts simple and easier to use.

QuickBooks will let you add, delete, or rearrange the Class list at any time. But like the Chart of Accounts list, it's best to set up at least the basic structure of the Class list before you begin entering transactions, then review it annually before entering transactions for each new accounting year. Putting some thought into how the Class list should be organized will result in more useful and useable enterprise information at year's end.

Guidelines for Setting Up Classes

The topics below present some points to observe when setting up classes in QuickBooks, to allow getting as much management information as possible out of your records.

Is Class Tracking Turned On?

You can only assign classes to transactions if the class tracking preference item is turned on. To check whether it is, open the Write Checks window (choose Activities|Write Checks) and look for a Class column. If you don't find one, turn on class tracking as follows:

1. Choose File|Preferences to open the Preferences window.

2. In the Preferences window, select the Accounting icon in the scroll box along the left side.

3. Your Preferences window may have a Company Preferences tab at the top. If so, click on it.

4. Select the "Use class tracking" check box.

5. Click OK to close the Preferences window.

♦ There are no right or wrong ways to set up classes, only different ways.

However, some class setups are rather inflexible and don't yield management information as easily as others.

Expect to redo parts of the Class list during your first year or two of using QuickBooks—consider it just part of your "tuition cost" for learning Quick-Books!

♦ Each class should usually represent a profit center or a cost center, or should be a "parent" class for a group of profit or cost centers.

PROFIT →
|
CENTERS
|
COST →

Profit centers are enterprises operated with the intent of making a profit. They normally correspond with the things you produce, like corn, soybeans, cotton, or feeder cattle.

Cost centers are enterprises which mainly exist to serve other (profit center) enterprises. A typical example would be a combine, which may not generate revenue on its own but exists to provide harvesting services to various profit centers.

Unless your class list is quite small, you will normally need at least two class levels; that is, "parent" classes and "child" or subclasses. Why? Some transactions—usually expenses—cannot reasonably be assigned to one specific enterprise, yet they can and should be identified with a group of related enterprises. Assigning these transactions to a parent-level class may get them closer to where they should be ultimately be assigned and will allow them to be more accurately allocated to the appropriate enterprises later.

Here's an example. Given this partial Class list:

Crops
    Grains
        Corn
        Soybeans
    Hay

- Repairs to a planter you use for planting both Corn and Soybeans are not a direct cost of either enterprise. Rather, they are an overhead expense which should be assigned to Grains, which encompasses both Corn and Soybeans.

- If you rotate between hay and grain crops on most fields, the cost of soil testing is an overhead expense most reasonably assigned to the Crops class. It is a cost shared by all crop enterprises.

What do you do with the costs that you accumulate for parent-level classes like Crops or Grains? To estimate production costs for specific crops you would allocate those costs among the child-level classes (Corn, Soybeans, Hay, etc.) on whatever basis seems reasonable.

◆ **Don't make class names serve more than one information goal.**

Instead, establish a different class level (i.e., use subclasses) for each type of information you want from the Class list. This idea is simpler than it sounds, and best explained by example.

The following class names serve two information objectives: each represents both an enterprise *and* a crop year.

Corn '97
Soybeans '97
Wheat '97

Assuming information is kept for multiple crop years, here's what the same Class list would look like after adding two more years of records:

Corn '97
Corn '98
Corn '99
Soybeans '97
Soybeans '98
Soybeans '99
Wheat '97
Wheat '98
Wheat '99

Though this setup would basically work, it presents at least two problems:

- Getting a report for all crops in any particular crop year requires filtering the report to include three different classes, such as Corn '97, Soybeans '97, and Wheat '97.

- Having so many class names beginning with the same characters makes QuickBooks' name lookup features less useful. When you begin typing in a form's Class field, QuickBooks tries to find the right class name by "looking ahead" as you type. When several class names begin with the same characters, you must type more of each name before QuickBooks finds the name you want.

Here's the same Class list, rearranged so that each class or subclass level relates to just one type of information:

> 1997
>   Corn
>   Soybeans
>   Wheat
> 1998
>   Corn
>   Soybeans
>   Wheat
> 1999
>   Corn
>   Soybeans
>   Wheat

This arrangement takes care of both problems mentioned above:

- Getting a report for all crops in any particular crop year only requires filtering the report on a single class, such as "1998" for the 1998 crop year.

- "Look-ahead" field filling works much more efficiently. You would only have to type 1998:S to have QuickBooks fill in the class name 1998:Soybeans.

- **The arrangement of class and subclass levels determines how easily you can get reports of specific class information.**

The basic rule is this: put the classes for which you will most often want reports at the upper levels of the class hierarchy. In a class identified as Crops:Grains:Wheat, Crops would be the highest-level Class; Wheat the lowest-level:

> Crops
>   Grains
>   Wheat
>   ...

Here are some examples of how different information goals might affect the arrangement of the same set of classes. In each case, generating reports based on the upper class levels is easiest, requiring the least report filtering. *Filtering* is the QuickBooks term for controlling or limiting which transactions are included in a report.

Goal: easy reporting by crop year—make the crop year the upper-level class.

```
1997
    Corn
        Jones farm
        Smith farm
    Soybeans
        Jones farm
        Smith farm
1998
    Corn
        Jones farm
        Smith farm
    Soybeans
        Jones farm
        Smith farm
1998
    Corn
        Jones farm
        Smith farm
    Soybeans
        Jones farm
        Smith farm
```

Goal: easy reporting by crop—make the crop the upper-level class).

```
Corn
    1997
        Jones farm
        Smith farm
    1998
        Jones farm
        Smith farm
Soybeans
    1997
        Jones farm
        Smith farm
    1998
```

    Jones farm
    Smith farm
Wheat
    1997
        Jones farm
        Smith farm
    1998
        Jones farm
        Smith farm

Goal: easy reporting by farm location—make the farm the upper-level class.

Jones farm
    Corn
        1997
        1998
    Soybeans
        1997
        1998
Smith farm
    Corn
        1997
        1998
    Soybeans
        1997
        1998

♦ **The number of ways to use classes is limited only by your imagination.**

Whenever you want to track information that doesn't relate to specific accounts, consider setting up classes to do the job. Here are a few more ideas about ways to use classes in your farm records.

You can use classes to keep track of almost any special category of income or expense. For instance, the following class structure would let you monitor herbicide cost differences in no-till and minimum tillage situations:

Corn
    Jones farm
        No-till
        Min-till
    Smith farm
        No-till
        Min-till

You can use classes to keep track of repair costs and other expenses for specific machines:

Equipment
    9500 JD Combine
    7120 Case IH
    Pickup
    Tandem Truck

If you use classes to identify operating costs for individual machines the ManagePLUS add-on, discussed in chapter 7, can provide reports of repair cost per hour or per acre.

Here's part of a class list which would provide detail for some forage enterprises:

Forage
  Hay
    Alfalfa
      Prime
      No. 1
      No. 2
      No. 3
      Common
    Clover
    Mixed Grass/Legume
    Grass
    Stover/Waste
    Straw
  Pasture
    Cool-season
    Warm-season

The Hay and Pasture subclasses would allow separating hay-related overhead expenses from pasture-related overhead. The Hay subclasses (Alfalfa, Clover, etc.) would allow keeping track of expenses that are specific to different hay crops, such as fertilizer, and would also provide separate reporting of hay sales for each class of hay. Finally, Alfalfa's subclasses (Prime, No. 1, No. 2 , etc.) would provide even more information about alfalfa hay sales by tracking the dollar value of each grade of alfalfa sold.

With the ManagePLUS add-on, discussed in chapter 7, you could also get a **report of the quantity sold** for each Hay class and subclass, including the average price received per ton or per bale.

Here's another example which shows how classes can be used to identify costs in specific areas. In this case, expenses associated with different parts of the beef cow enterprise are identified with subclasses.

> **Cattle**
>   **Beef Cows**
>   **Neonatal Calves**
>   **Pre-Weaning**
>   **Cows**
>   **Backgrounding**
>   **Replacement Heifers**

Neonatal Calves identifies costs associated with getting calves started during their first week of life (antibiotics, veterinary fees, colostrum supplements, etc.). Pre-Weaning identifies calf costs from a week of age through weaning (deworming, medications, creep feed, etc.). Cows identifies expenses specific to the cows themselves (feed, minerals, medications, etc.). The Backgrounding enterprise tracks calf costs after weaning and during overwintering or a summer grazing period. Replacement Heifers tracks costs specific to heifers being raised as cow herd replacements (veterinary fees, artificial insemination costs, etc.).

You can also use classes just as a way to tag certain transactions, to make them easier to find and group together to get a total. For example, suppose several farm buildings have received wind damage. Your insurance agent has agreed to let you make some of the repairs and hire a carpenter to make the rest. The repairs will be paid for by the farm business, and the insurance company will provide a reimbursement based on the total repair costs. An easy way to identify all of the repairs and get a report which totals them is to tag all of the wind damage-related repair costs with a class created specifically for that purpose:

> **Wind Damage '99**

How to Set Up New Classes

Like accounts, you can set up new classes from any QuickBooks form which has a Class field or by using commands in the Class List window, as the following steps describe.

1. **Open the Class List window.**

Choosing Lists | Classes from the main menu is one way to do this.

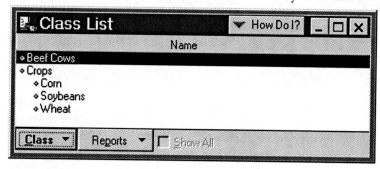

2. **Click on the Class button at the bottom of the window.**

A pop-up menu of commands for working with classes will be displayed.

3. **Click on New in the pop-up menu.**

QuickBooks will open the New Class window. (This same window will open if you choose to add a new class while entering a transaction.)

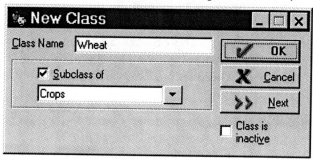

4. **Fill in information about the new class.**

Here are descriptions of the New Class window's fields.

Class Name
This is the name that will appear in the Class List window and also the one you will type or select when entering a transaction.

Subclass of
To make your new class a subclass of another, click on the check box and select the name of the parent class in the box beneath it.

5. **Click OK or Next to save the new class in the class list.**

Clicking OK saves the class and closes the New Class window. Clicking Next keeps the New Class window open for entering another class.

Items: Identifiers for the Things You Buy and Sell

Items are identifiers you can set up in QuickBooks to represent things you buy and sell. Depending on how you like to enter income and expense transactions, they can make transaction entry faster and easier for things you buy or sell frequently.

Suppose you are entering a corn sale transaction on the Cash Sales form. If you previously have set up a Corn item you can simply choose it from the Items list. Without any more effort on your part, QuickBooks will fill in the transaction description assigned to the Corn item and will post the income to the appropriate account.

Using items also lets you enter quantities for the things you buy and sell, such as the number of bushels of corn sold or the gallons of fuel purchased. Sale and purchase quantities can only be entered on QuickBooks forms which permit using items: Cash Sales, Checks, Bills, and Invoices. Forms which only allow selecting accounts do not support entry of sale and purchase quantities.

Is there a way to enter quantities without using items?

Yes. The ManagePLUS add-on, discussed in chapter 7, allows entering quantity information in most QuickBooks forms, including forms where items cannot be used such as the Deposit form and the General Journal.

Some QuickBooks forms such as Checks and Bills let you make entries either with items or by selecting accounts, whichever you choose. On those forms you click on the appropriate tab, either Expenses or Items (shown below), to pick the entry method you want to use.

Expenses	$0.00	Items	$0.00

Using items in QuickBooks 4.0 and earlier versions requires that you first turn on QuickBooks' inventory features, even if you don't plan to track inventories. You can tell whether inventories are turned on by opening the Write Checks window and looking at the lower half of it. Both an Expenses tab and an Items tab will be present if the inventory features are turned on. If not, here's how to turn them on:

1. From the QuickBooks main menu choose Preferences|Inventory/PO's.

    The Inventory/Purchase Order Preferences window will be displayed.

2. Select the "Inventory and purchase orders are active" check box.

3. Click OK.

Guidelines for Setting Up Items

Do You Need Items?

It is possible to operate QuickBooks entirely without using items, but most people use at least a few of them. Items are required if you want to do any of the following:

♦ Enter quantity information in your transactions.

♦ Use QuickBooks' inventory features.

♦ Enter sales on the Cash Sales form.

♦ Enter sales on account (using the Invoice form), and keep track of customer accounts.

Even if none of these activities are important to you, you may find that you prefer using items to enter certain types of transactions, particularly sales.

Choosing Appropriate Item Types

QuickBooks defines different item types for different purposes. Choosing the appropriate item type is the most important step in setting up a new item. An item's type determines how the item works; how transactions in which the item is used will post income, expense, or inventory changes to your accounts.. A few of the more important item types are discussed below. Consult the QuickBooks Help system for information about the other types.

♦ **Non-inventory Part**

Non-inventory Part items are used for entering purchases or sales and normally only record income or expense. Unlike Inventory Part items, they *do not* record changes in inventory quantities on hand. So only assign the Non-inventory Part type to items which are *not* used to represent things placed in inventory.

In a typical farm business most items will be of the Non-inventory Part type. It is used more often than the Inventory Part type only because of shortcomings in QuickBooks' inventory system which prevent it from properly handling most farm inventories. However, there are some work-arounds which allow using the inventory system to keep track of resale livestock and other resale items (page 199), and also inventories of farm production (page 187).

You may set up Non-inventory Part items for any type of farm production you sell, any input you purchase, or anything else that involves paying or receiving funds. Here are some examples:

> Corn
> Soybeans
> Wheat
> Raised Calves
> Stocker Cattle
> Milk
> Lime
> Anhd. Ammonia
> N (units)
> P (units)
> K (units)
> Corn seed (units)
> Soybean meal (ton)
> Soybean seed (50 lb)
> Soybean seed (bu.)

- **Service**

Unless you are using the time billing features of QuickBooks Pro (which are used mostly by CPAs, attorneys, and others who bill for their services on the basis of time spent on a job), QuickBooks treats items of the Service type exactly like items of the Non-inventory Part type. There are no special reasons for using the Service item type in a typical farm business, except that it gives you an additional way to organize items in the Items List window (where QuickBooks groups items by item type.)

A good use of the Service item type is for items which identify services you perform for others (such as custom harvesting) or services you purchase from others (custom application, soil testing, etc.). Here are some examples:

> Custom Combining
> Custom Hay Baling
> Herbicide Application
> Fertilizer/Lime Application
> Custom Soil Testing
> GPS Data Conversion

- **Inventory Part**

QuickBooks only makes the Inventory Part item type available if you enable inventory tracking, by selecting the "Inventory and purchase orders are active" check box on the Purchases & Vendors panel of the File | Preferences window.

Like the Non-inventory Part type, items of this type are used in sale or purchase transactions and normally record income or expense. The difference is that when an Inventory Part item is used in a transaction, an inventory quantity (and also an inventory asset account) are affected. This is the only item type you may use for tracking purchases and sales of inventoried goods.

QuickBooks' inventory system was designed for retail businesses, and most farm inventories (grain, market livestock, feed, seed, fuel, etc.) don't fit the retail goods inventory model. Consequently, most farm business users of QuickBooks avoid the inventory system entirely except for the few types of farm inventory that fit (or can be made to fit) the retail goods model. These are primarily resale livestock and other resale items (page 199), and inventories of farm production (page 187)—but *not* inventories of farm inputs. Before setting up anything you buy or sell as an Inventory Part item, be certain you understand the requirements of using the QuickBooks inventory system.

For examples of Inventory Part items for a farm, see the topics which discuss resale livestock (page 199) and farm production inventories (page 187).

◆ **Discount**

QuickBooks only permits using Discount type items on the Cash Sales and Invoice forms. On those forms, a Discount type item produces in a *negative* dollar amount on the transaction line(s) where it is used. Discount items are handy for entering sales deductions, like commissions and yardage deducted from a livestock sale or a dairy cooperative's deductions from a milk check. Here are some examples:

> **Commission Ded.**
> **Yardage Ded.**
> **Insurance Ded.**
> **Hauling Ded.**
> **Advertising Ded.**
> **Capital Retain Ded.**

Set Up Separate Items for Different Quantity Units

If you use items because they allow entering quantities for the things you buy and sell, there's another thing to know about setting up items. To keep from totaling "apples and oranges" quantities, you need a *separate* item for each different quantity unit in which the item is bought or sold. Some typical examples will illustrate this idea better.

Soybean seed is commonly sold in both 50-pound units and bushel (60-pound) units. Unless you set up separate items for both unit sizes, any total for the number of soybean seed units purchased may be meaningless.

Many herbicides are sold in different concentrations and different formulations. Over the years the atrazine product has been available as a liquid formulation (4L), a dry formulation (80W), and a dry flowable (90DF), and probably some others. When the same basic product is available in different formulations or is purchased in different units (like ounces vs. pounds vs. gallons), you need a separate item for each one. Quantity totals for a "catch-all" item will be meaningless if it is used for purchases of different concentrations or quantity units.

Feedstuffs are often available with different protein and/or energy levels. For example, both 44% protein and 48% protein soybean oil meal are available. If your supplier sometimes has one or the other, you need separate items for recording purchases of each type. Otherwise, the total quantity of soybean oil meal purchased for the year won't give you specific information on the amount of protein that has actually been purchased.

When you have several different items representing essentially the same item—but different quantity units—how do you tell them apart? QuickBooks requires that every item have a unique name. Why not use that requirement to your advantage, by including something in the item name to indicate the quantity unit? Here are examples based on the discussion above, which show how different products or units of measure can be expressed in item names:

SB Seed (50 lb.)
SB Seed (bu.)

Atrazine 80W (lb.)
Atrazine 90DF (lb.)
Atrazine 4L (gal.)

SBOM 44%
SBOM 48%

How to Set Up an Item

Like accounts and classes, you can set up new items from any QuickBooks form which has an Item field or by using commands in the Item List window, as the following steps describe.

1. **Open the Item List window.**

Choosing Lists|Items from the main menu is one way to do this.

Here's how the Item List window might look with a few items added:

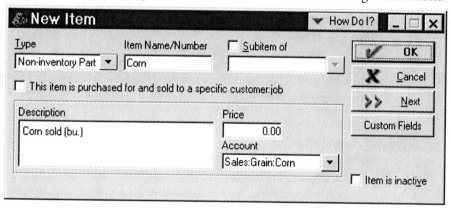

Item List

Name	Description	Type	Account	Price
◇ Hfrs-Feeder	Feeder heifers sol...	Non-inventory Part	Sales:Livestock:Cattle	0.00
◇ Soybeans	Soybeans sold (bu.)	Non-inventory Part	Sales:Grain:Soybeans	0.00
◇ Strs-Feeder	Feeder steers sold...	Non-inventory Part	Sales:Livestock:Cattle	0.00

2. **Click on the Item button at the bottom of the window.**

A pop-up menu of commands for working with items will be displayed.

3. **Click on New in the pop-up menu.**

QuickBooks will open the New Item window. This is the same window QuickBooks opens if you choose to add an item while entering a transaction.

4. **Fill in information about the new item.**

Here are descriptions of the New Item window's fields. (Additional fields are displayed for some items, such as Inventory Part items. Consult the Quick-Books Help system for information on those fields.)

Item Type

Choosing the appropriate item type is the most important step in creating a new item. Item types were discussed earlier in this section.

Item Name/Number

Enter the name you want assign to the item. This is the name you'll see when picking an item from the drop-down list of items on check, bill, invoice, or cash sales form.

Tips for naming items

Compared to Quicken, QuickBooks gives you a lot more space for entering information, so abbreviating names in most lists isn't necessary. However, keeping item names reasonably short is still a good idea.

If you normally select items in transaction entries by typing (as opposed to picking them from the Items list), less typing will be necessary if the first few characters of item names are mostly unique.

For example, suppose you have items named Feeder Steers and Feeder Heifers. To select either item you would have to type "Feeder " before typing the "S" or "H" that would uniquely identify the desired item. "Fdr Steers" and "Fdr Heifers" would be better names because you'd only have to type "Fdr S" to select the steers item or "Fdr H" to select the heifers item. Even better names would be "Strs-Feeder" and "Hfrs-Feeder". For those you would only have to type an "S" or an "H" to select the desired item.

You can change an item's name at any time, using commands in the Item List window. QuickBooks will automatically make the necessary changes so that all transactions using the item will show the new name.

Description

When you select an item in a sales or expense form, QuickBooks automatically displays the item's description on that line of the form. Though you can edit the description, it's a good idea to set up items with descriptions that seldom need editing. And if you later decide the description is not quite right, you can always come back to the Items list and edit the item to change the description.

When creating item descriptions it is helpful to include the appropriate purchase/sale unit (lbs., gal., bu., etc.). That way, when QuickBooks copies an item description onto a sales or expense form it will serve as a reminder of the appropriate unit of measure for the item. You won't be left wondering whether you should enter a livestock sale quantity in pounds or in hundredweights, for instance.

Price

Generally speaking, you should only enter something in the Price field for items you sell, and only if those items are normally sold at a price you determine. Otherwise, leave the Price field blank or 0 (because you'll always have to enter a price for the item anyway, wherever it is used).

If you added a Custom Combining item to use for entering custom harvesting income, putting a value in the Price field would be appropriate. But you shouldn't do so for a Wheat Sales item. Why? Wheat will be sold at many different prices, so having QuickBooks pre-enter a price when you record a wheat sale could be more of an aggravation than a benefit.

Account

You must associate an Account with each item so QuickBooks will know how to post transaction amounts involving the item.

For Inventory Part items you must select three Accounts: an asset account, where the item's value will be posted when you purchase the item; a cost of goods sold account, usually Cost of Goods Sold; and an income account, where income will be posted when you sell the item. Work-arounds for handling farm inventories sometimes involve non-standard account selections for Inventory Part items. (See the grain inventory topic on page 187 for examples.)

For Non-inventory Part and Service item types you will normally associate a single income or expense account with the item.

Let's consider how this works. Suppose you set up a Non-inventory Part item called Seed Corn and associate it with the Seed Expense account. When you select the Seed Corn item to enter a seed purchase, the dollar amount of the purchase will automatically be posted to the Seed Expense account, increasing its balance. If you return some of the seed and need to enter a refund you again select the Seed Corn item, but this time the dollar amount is negative, which decreases the Seed Expense account's balance.

Once in a while you may need to set up two different items representing the same commodity, because you want to post purchases and sales of the commodity to different accounts. Here's an example. You might set up an item named Corn, associate it with the Corn Sales account, and use it for entering corn sales transactions. But if part of the time you need to purchase corn for livestock feed, you could set up another item called Corn-Feed and associate it with the Feed Expense account.

5. **Click OK or Next to save the new item in the Items list.**

Clicking OK saves the item and closes the New Item window. Clicking Next keeps the New Item window open for entering another new item.

Name Lists: Customers, Vendors, and More

QuickBooks keeps four name lists identifying the people and companies you do business with: Customers (Customers:Jobs), Vendors, Employees, and Other Names.

Customers includes any person or business who pays money to the farm business. That's broader than the usual definition of "customers"—a better name might have been Payers. This list includes the grain elevator, livestock auction, and other businesses and individuals who buy your grain, livestock, or other production. But it also includes every other person or business who pays money to the farm business for any reason.

You also have the option of setting up one or more Job names for each customer. Referred to as Customer:Job names, these are most often used in service-related businesses to separately track expenses for, and bill for, various jobs undertaken for a customer. A typical example would be a carpenter working on different parts of a house remodeling job. The Customer:Job feature is sometimes useful in farm accounting situations, particularly if the farm's sales involve providing a service, such as a sod farm which does residential and commercial sod laying.

Vendors includes any person or business to whom the farm business pays money. Besides the usual suppliers you might think of—the local crop service, feed dealer, and machinery dealer—this list includes your bank; credit card companies; county, state, and federal tax collection agencies; and many others.

Employees includes any person to whom the farm business has issued a paycheck using QuickBooks payroll.

Other Names is for names which aren't in the Customers, Vendors, or Employees lists. QuickBooks adds names to the Other Names list when importing records from Quicken or from old versions of QuickBooks if it cannot determine whether a name should be a customer or a vendor. You may move Other Names to any of the other name lists (there's a Change Type button in the Edit Other Names window), but they cannot be moved back. In other words, Customers, Vendors, and Employees cannot be moved to the Other Names list.

Generally you should not add names to the Other Names list, because the names in it can't be selected in QuickBooks forms where a customer or vendor name is required. However, you may use the Other Names list for special activities such as keeping a small mailing list.

Why Spend Time Entering Names?

Most QuickBooks forms require that you enter a customer or vendor name. To keep from having to type in lots of names you could set up generic names like Cash Sale and Cash Purchase, and use them most of the time. You would avoid a lot of typing...but you'd also miss all of the benefits that come from having detailed name lists!

Adding customer and vendor names is so easy, even while entering a transaction, that you should eventually try to get all of your regular customers and vendors entered in those lists. Here are some good reasons why:

- **Identifying the customer or vendor by name in a transaction adds useful detail.**

   When looking through prior transactions you'll be able to match a specific person or business with each transaction. If all of your transactions identified the customer name as "Cash Sale", figuring out a billing or payment problem might take forever.

- **Identifying customers and vendors by name makes it easy to search for transactions.**

   QuickBooks' Find command can search for all transactions containing a specific customer or vendor name.

- **Supplying a mailing address when you add a name to the Vendors list lets you mail out checks in windowed envelopes.**

   When you select a vendor name on the Checks form, if the vendor has a mailing address QuickBooks adds it to the check. When the check is printed, it will be ready to stuff in a windowed envelope—no hand addressing needed!

- **You can do a customer mailing if you've entered mailing addresses for your customers.**

   Do you have regular hay customers, or a U-pick pumpkin patch? Quick-Books can print mailing labels from the Customer list, making it easy to send out a flyer or letter to your customers.

- **Having complete Customer and Vendor lists makes using QuickBooks easier.**

   The work you put into getting customers and vendors entered into Quick-Books will eventually pay off: when you enter a check or a bill, you need only type a few characters to select the appropriate vendor and have the vendor's full name and address added to the check, ready for printing.

Setting up "generic" customers and vendors

Though you should use actual customer and vendor names in most transactions, generic names do have a place. You can set up Cash Sale and Cash Purchase as customer and vendor names, respectively, and use them in transactions when you don't know the customer's or vendor's name. They're also handy for entering one-time sales or purchases—they let you avoid cluttering the name lists with names you expect to use only once.

Another variation is to set up the names One Time Customer and One Time Vendor. Note that these names begin with a zero, not an "O". Because they begin with a number instead of a letter, QuickBooks will sort these names to the top of the Customer and Vendor lists. And they're easy to select when filling in the Customer or Vendor field on a form: just type "0" (zero)!

If spelling the names with a zero at the front seems confusing, try using a "1" (one) instead, like this: 1 Time Customer, and 1 Time Vendor.

Adding Customer, Vendor, and Employee Names

There's no need to spend hours entering names before you start entering transactions in QuickBooks. Just enter new names as needed, while you enter transactions—adding new names then is usually easiest. That's when you are most likely to have the necessary information close at hand. The person's or business' name, address, telephone number, and other information will often be right in front of you, printed on the check, invoice, or statement you are entering.

Here are the steps for adding new names as you enter transactions.

1. **Type a name in the Customer or Vendor field of the form where you are entering a transaction.**

   QuickBooks will search the name list as you type, and highlight the name if it is already present in the list. If the name is not in the list, when you attempt to move to another field on the form, QuickBooks will ask how to handle the new name:

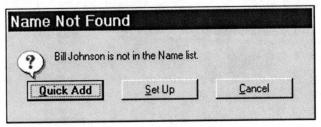

2. **Choose how to handle the new name.**

- Click on Quick Add if you want to immediately add the name without entering other information like the person's address or telephone number. You'll typically use Quick Add when you don't have any information other than a name, or when you feel it's not important to keep other information about the customer or vendor.

- Click on Set Up when you want to enter more than just a name. The New Customer or New Vendor window will open, and you can fill in the address, telephone number, customer or vendor type, etc., as described farther below.

- Click on Cancel if you mistyped the name and want to retype it or you want to select a different name.

If you click on Set Up, a window will open to let you enter details about the new name, such as the New Vendor window shown below. (The windows for the other name lists are nearly identical.)

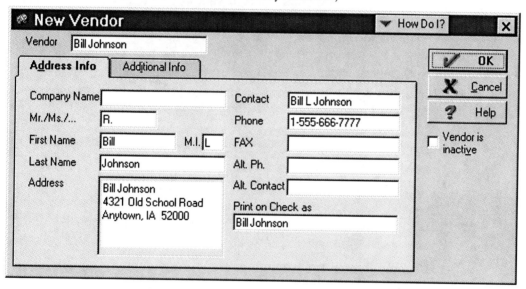

3. **Fill in details about the person or business you are adding to the name list.**

Here are comments about a few of the most important fields. For information about other fields, including those on the Additional Info tab, consult the QuickBooks Help system.

Vendor, Customer, Employee, or Name

The name of this field depends on the list in which it appears, but it is always the first field at the top of the window (the Vendor field in the window above). Whatever you enter here is what QuickBooks will display in the name list and what you will see when selecting a name in a form.

When you enter a new name QuickBooks places it in the name list alphabetically according to this field. If you enter "Al Ziegler" the name will appear toward the beginning of the list. But if you enter "Ziegler, Al", it will appear toward the end. We are all accustomed to seeing names ordered alphabetically in telephone directories and such, so working with names is usually easiest if they are entered in "last name, first name" order—i.e., "Ziegler, Al".

The name you enter in this field must be unique across all name lists. What if the same person needs to be in more than one list? If Al Ziegler is both a customer and a vendor, enter his name a bit differently in both lists. You can use a middle initial, include a period or a comma, add a digit...anything that will make the names slightly different. The address, telephone number, and all other information, however, can be identical in both lists.

Address

The address field is the same for each name list, but it has an additional purpose in the Vendors list. Whatever you enter in the Address field will be printed in the address area of checks you print from QuickBooks. If you want to mail checks in windowed envelopes, supply a complete name and postal address here.

Print on check as (Vendors list only)

The only concern with entering vendor names in "last name, first name" order is that you probably don't want names printed that way on checks—you don't want the payee printed as "Ziegler, Al" on checks to Al Ziegler. To get around the problem type "Al Ziegler" in the Print on Check as field.

Note that the name order isn't important if Al Ziegler is a Customer. You can type "Al Ziegler" in the Company Name field and the Address field. These fields get printed on any sales forms Al might see, but the Customer field (where you typed "Ziegler, Al") does not.

4. **Click OK or Next to save the new name in the name list.**

Clicking OK saves the name and closes the name entry window. Clicking Next keeps the name entry window open for entering another new name.

CHAPTER 4

Getting <u>Real</u> Work Done with QuickBooks!

QuickBooks gives you so many different ways to do things and offers so many potentially valuable features, it's hard to know which ones to use. Which features will *really* save time? Or make learning and using Quick-Books easier? Or provide the most management information?

The purpose of this chapter is to help you answer these questions quickly. It offers a number of tips and recommends specific QuickBooks features and accounting practices every QuickBooks user should consider.

New Users: Try the Navigator

The Navigator was introduced in QuickBooks 5.0 to give you a more graphical view of, and access to, QuickBooks features.

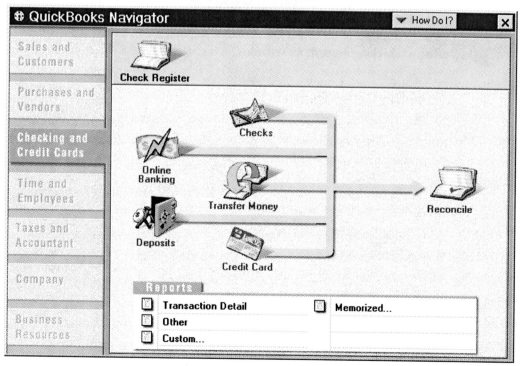

To turn the Navigator on or to move it in front of other open windows, click on the QB Navigator button in the upper right corner of the QuickBooks main window. To turn the Navigator off, click on the close button (the small square button marked with an **X**) in the upper right corner of the Navigator window.

Why Use the Navigator?

The biggest hurdle for most new QuickBooks users is learning how the many different forms and other features work together, and the proper sequence for using them. The Navigator was designed to help with this. It shows the appropriate QuickBooks forms to use for jobs like working with a checking account, as well as the usual sequence of activities, displayed as flowcharts. Flowcharts for different jobs are available by clicking on the topics displayed along the left side of the Navigator window.

Forms are represented by icons (small pictures) within Navigator, and you can click on an icon to go directly to the desired form. If you are new to Quick-Books you may find this easier than working with the program's menus. In the Navigator window example above you could open the Write Checks window by clicking on the Checks icon. Or, you could choose Activities|Write Checks from QuickBooks' main menu.

Each Navigator window also gathers together icons for the lists, actions, and reports that relate to a particular activity. For example, you don't have to search through multiple levels of the QuickBooks' Reports menu to find reports that relate to checking account transactions, you just click on the appropriate icon.

What Navigator Doesn't Tell You...

In some places the Navigator suggests two or more different forms to choose from. But it doesn't tell you which form to use in different situations or whether your choice makes any difference. For example, the flowchart on the Sales and Customers tab shows that you can enter sales on either an invoice or the cash sales form, but leaves you with the responsibility of knowing which one you *should* use.

Later chapters tell which forms to use in various situations. The QuickBooks Help system also does a good job of describing the purpose of each form, although applying the information to farm accounting situations isn't always easy.

Keep Your Records Current!

The accounting job is always easier if you enter bills soon after you receive them, checks soon after they are written, and deposits when you deposit the funds. Staying current is important in any accounting system, computerized or manual, but it's even more important with QuickBooks. QuickBooks is designed so that keeping your records current is a key ingredient in reducing the amount of time you spend on record keeping.

If you are coming to QuickBooks from a manual ledger system or a less-modern computer program you may not fully appreciate the benefits of keeping your accounting entries up to date. In those systems the main encouragement for keeping your records current is the threat of having a huge backlog of bookkeeping work to do to catch up, but with QuickBooks there are real benefits of staying current. You can:

♦ Avoid spending much time paying bills when you are busiest with other work

♦ Have an up-to-date, calendarized schedule for paying bills as they come due

♦ Reconcile your checkbook and credit cards the same day you receive your statement

♦ Always know your checking account (also credit card) balance

The point to make here is that you will only experience these benefits if you make the effort to keep your QuickBooks accounting entries current!

Use the Reminders Window

Get into the habit of using the Reminders window as your "command center" to stay aware of accounting chores you need to do.

The Reminders window (shown on the following page) is like an automated notepad that QuickBooks maintains, which lists current and pending bookkeeping activities such as bill due dates and amounts, important dates, and to-do notes. It can list the bills coming due over the next thirty days, show checks you've entered but not yet printed, and a lot more.

Go to transactions directly from the Reminders window by double-clicking any detail line in the window, such as the individual bills listed in the example below. The underlying transaction will be opened for viewing and/or editing.

Due Date	Description	Amount
	To Do Notes	
03/01/1999	Get land rental contract to Bill for signature	
	Money to Deposit	2,510.00
	Bills to Pay	-3,999.00
03/12/1998	Agri Services, Inc.	-3,357.00
03/20/1998	Robert A. Jones, CPA	-210.00
01/22/1999	Unified Electric Cooperative	-110.00
01/22/1999	Abel's Repair	-322.00
	Overdue Invoices	12.50
	Sales Receipts to Print	2,537.50
	Memorized Transactions Due	0.00

Reminders — How Do I?

Normal   Summary   Detail

To benefit the most from the Reminders window you need to see it often. QuickBooks even has a preference setting (see page 24) that causes the Reminders window to be displayed each time a company file is opened.

Manage Your Bill Payments with QuickBooks

In some farming operations, QuickBooks' features for managing bills—keeping track of due dates and amounts, forecasting cash requirements, paying bills on time, and even printing the checks—are considered to be the biggest benefit of using the software. You are likely to benefit significantly from using QuickBooks' bill management features, *even if it requires a change in your accounting habits.*

Bills is the term QuickBooks uses for what accountants more often call accounts payable—the amounts you owe for purchases on credit.

Here are three main benefits of letting QuickBooks help manage your bills:

1. It *really does* save time compared to doing the job other ways.

2. It reduce the time demanded by bookkeeping, at those times of year when you are busiest with other things like calving, planting, or harvest.

3. It will help you form habits that work almost magically to keep your accounting records up to date.

For specifics on using QuickBooks' bill management features, see "Working with Bills" in chapter 5.

Use QuickFill to Speed Transaction Entry

Most of the work of entering transactions involves selecting accounts, classes, names, etc. from various lists. The QuickFill feature greatly speeds up selecting things from lists.

A standard way of selecting from a list is to use the mouse. In the field where you want to select something from a list you click a down arrow, causing the list to be displayed, then use the mouse and scroll bars to scroll through the list to find and select the desired item. This method is fine, especially while you're learning QuickBooks or when you are unsure of how the item's name is spelled.

But QuickFill is faster. It lets you find a list entry by typing just the first few letters of its name. As you type each character, QuickFill searches the list and displays the closest match. For example, suppose part of your chart of accounts looks like this:

Seed
Supplies
Taxes
Utilities
    Electricity
    Water

And suppose you're entering a check for supplies. To select the Supplies account in the check's account field, just start typing...by the time you've typed "Su" QuickBooks will have displayed "Supplies" in the field, and you can move on to the next field.

QuickFill can also quickly locate subaccounts and subclasses, especially if their names begin with characters that are unique throughout the list. Given the accounts listed above, typing "W" in an account field would immediately fill the field with "Utilities:Water", because no other account or subaccount begins with "W".

QuickFill makes it easy to select a string of account and subaccount names. To enter "Utilities:Electricity" you could type "U:E". Typing the "U" finds the Utilities account, typing ":" tells QuickFill to continue the rest of its search among the subaccounts of Utilities, and typing the "E" finds the Electricity subaccount.

Let QuickBooks Help You Reconcile Your Checkbook and Credit Cards

QuickBooks makes checkbook and credit card reconciliation easy and, in the process, can save you a lot of time. When a bank statement arrives all you need to do is (1) open the Reconcile window (by choosing Activities|Reconcile from the main menu), (2) select the checks and deposits which have cleared the bank, and (3) verify that the cleared balance matches the balance shown on the bank statement.

The key to making reconciliation most useful is to keep your records current; that is, keep checks, deposits, and other QuickBooks entries up to date. If you do, you'll be ready to reconcile your account as soon as the bank or credit card statement arrives. That will help you catch errors before the transactions get "cold" and the details are difficult to remember. It will also verify whether your QuickBooks account balances match the balances in your bank account or credit card account.

For details on reconciling accounts see the chapter 5 topics "Working with Checking Accounts" and "Working with Credit Cards", or consult the Quick-Books Help system.

Use Memorized Transactions?

QuickBooks can save you a lot of typing by "memorizing" the details of transactions you enter often, storing them in the Memorized Transactions list for reuse later. Some QuickBooks users depend heavily on memorized transactions; others find them less useful. Whether *you* decide to use them probably will depend a lot on the types of transactions you enter.

Memorizing transactions is most useful for those transactions you enter over and over, in almost identical form each time except for details such as the date and dollar amount. Examples of good candidates for memorizing would be the utility checks you write each month for electricity, telephone service, and rural water, or any other transactions which you usually assign to the same set of accounts and classes, such as a milk check.

You'll find details about memorized transactions in the topic Using Memorized Transactions in chapter 9, but here are the basics:

♦ **To create a memorized transaction:** (1) open the form where the transaction will be entered, such as Checks or Bills, (2) create a model transaction by filling in just the fields you want to memorize, then (3) memorize the transaction by choosing Edit|Memorize *transtype*, where *transtype* names the type of transaction you are memorizing.

♦ **To use a memorized transaction:** (1) open the Memorized Transaction List window (choose Lists|Memorized Transactions), (2) select the transaction you want to use, (3) click on the Enter Transaction button in the Memorized Transaction List window to create a new transaction entry, (4) add or change information in the new transaction entry, as needed, then (5) click on the OK or Next button to save the new transaction.

QuickBooks has a feature called **Automatic recall**, which *automatically* **provides most of the benefits of memorized transactions.** It works like this: when you select a customer or vendor name in a QuickBooks form, Automatic recall will automatically fill the rest of the form with details of your most recent transaction involving that customer or vendor. Then all you need to do is change any fields that need to be different for the current transaction, and save the transaction.

Automatic recall is a preference setting you can turn on. Choose File| Preferences, click on the General icon, then select the "Automatically recall last transaction for this name" check box.

Let QuickBooks Print Your Checks!

If you asked "Should I let QuickBooks print my checks?" the answer would be simple: "YES! YES! YES!" There are just three criteria you should meet:

1. **You want to save time.**

   Especially if you pay several bills at a time or generate a lot of checks, letting QuickBooks handle the details of check printing is a real time saver.

2. **You want help keeping your accounting records up to date.**

   When you hand write a check, the job of entering it later into QuickBooks is still left to do. But when you let QuickBooks print a check you are *done* with the transaction. Prior to printing it, you will have entered the check along with all of the necessary accounting details.

3. **You have a printer which feeds paper as single sheets.**

   Only a few years ago nearly everyone used a continuous-feed printer, the kind that uses one long, continuous ribbon of paper with perforations for tearing it into separate sheets. At that time, most of the arguments against computer-printed checks focused on the time, effort, and paper wasted when switching between plain paper and check forms in the printer.

Most modern printers are sheet-fed, meaning the paper feeds in one sheet at a time. With a sheet-fed printer if you want to print, say, three checks, you just place three blank check forms in the paper bin, and print. There's no wastage of check forms due to misaligned printing, and no need to remove the plain paper to use the check forms.

Printing checks with a continuous-feed printer is only a time and paper waster if you have to switch from paper to check forms every time you want to print checks. **If you have two printers** the problem disappears, even if both are continuous-feed printers. You can leave plain paper loaded in one printer and check forms loaded in the other, ready for printing at any time.

If you buy a new printer while your old printer is still in good condition, consider keeping it just for printing checks. All you'll need is another printer cable and possibly a second printer interface installed in your computer, for a total cost of $25 or less.

Answers To Some of Your Check Printing Questions

Here are answers to some other questions you may have about printing checks with QuickBooks.

+ **Can I still hand write some checks?**

   Certainly. Hand-written and computer-printed checks are entered into Quick-Books in exactly the same way, except that you type in the check number for a hand-written check but let QuickBooks fill in the check number for computer-printed checks. (Selecting the "To be printed" check box causes Quick-Books to supply the check number when the check is actually printed.)

   And of course, hand written checks are entered after you've written them, while the ones you print with QuickBooks are entered before printing.

+ **The check numbers on my hand-written checks begin at a different number than my computer checks. Will this put my checks out of order in QuickBooks' check register?**

   No. QuickBooks sorts the check register in date order. (In QuickBooks 6.0 and later versions you can choose the sort order for any Register window.)

+ **What if some my hand-written checks and computer checks have the same check numbers?**

   It's better if the two sets of checks are numbered differently. Besides eliminating some confusion, it lets QuickBooks watch for errors such as entering the same check twice. When you order computer checks, ask for a starting check number that is much higher (or much lower) than the numbers on your hand-written checks.

If some of your check numbers are the same, choose File|Preferences and then click on the Checking icon. Turn off the "Warn about duplicate check numbers" option.

- ◆ **Are pre-printed computer check forms expensive?**

Computer checks cost about the same or slightly more than other preprinted business checks ordered from your bank. Any additional cost is easily paid for by the time savings and convenience of printing checks with QuickBooks.

You may be able to order preprinted computer checks through your bank or through one of the many companies which specializes in pre-printed computer forms. See Appendix B for sources of check forms compatible with QuickBooks.

- ◆ **What about windowed envelopes?**

When you order checks don't forget to also order windowed envelopes designed to fit them! Windowed envelopes may seem expensive at first, but they soon pay for themselves in time savings. If you make a habit of entering addresses for new vendors you add to the Vendors list, every check you print from QuickBooks will be addressed and ready for mailing. Just stuff it in a windowed envelope, apply postage, and mail!

Use QuickBooks' Inventory Features Carefully

The inventory system in QuickBooks was designed for handling inventories of items bought for retail sale. Most farm inventories—grain, hay, fruits, vegetables, market livestock—are *produced*, not purchased, and would be better handled by an inventory system designed for manufacturing. But QuickBooks' inventory system cannot easily be made to work for the inventories of a manufacturing company; nor is there a practical way to make it work for most farm inventories.

This is a sensitive subject for both new and experienced QuickBooks users in agriculture.

If you are new to QuickBooks, maybe you purchased the program partly *because of* its inventory features, expecting they would make it easier to print a correct farm balance sheet at any time. Learning that the inventory system won't work as you had planned may be disappointing.

If you are an experienced QuickBooks user you may have tried several different approaches to using the inventory system. If so, you may be part of an elite club:

avid QuickBooks users resigned to the idea that there is no practical solution for the inventory system's shortcomings.

This is not to say you shouldn't use the inventory system in some cases. Here are a couple examples:

◆ Obviously, if part of your income comes from reselling items you have purchased, consider using the inventory system to keep track of them. One example is livestock purchased for resale, discussed beginning on page 199.

◆ You may manage farm production (stored grain, etc.) as QuickBooks inventories if you are willing to use the slightly non-standard techniques described on page 187.

Learn to Use Forms and Registers

Understanding how QuickBooks forms and registers are related will make you feel like an "old hand" at QuickBooks in no time at all. It will give you quick access to all of your transactions, including ones that may be weeks, months, or years old, and will make it easier for you to locate specific transactions.

Forms are the windows within QuickBooks where you enter and edit most transactions. QuickBooks forms mimic the appearance of real-world paper forms such as checks, deposits, bills, or invoices. Registers are windows which show all of the transactions entered in a particular account.

The important thing to understand about forms and registers is how they are related. Think of a form as a window (the real-world kind) into a register. A form shows one transaction from an account but a register lists all transactions from the account. You can switch back and forth between them at will, and can even have both open at the same time.

Here is a form you'll use a lot, the Checks form:

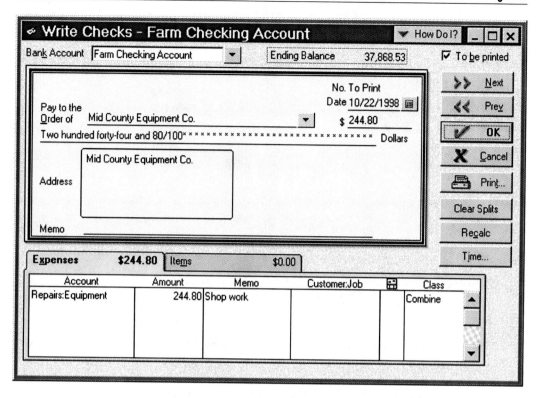

And here is the Register window for the same checking account. Note that the last line displays the same check as in the Checks example above.

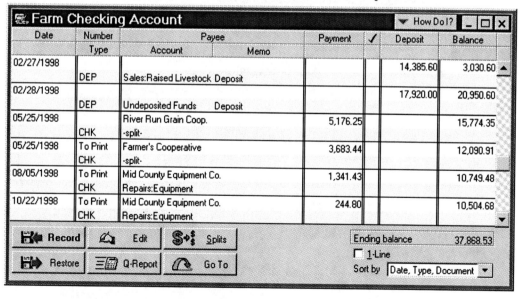

Registers are much more than just a listing of transactions. You can enter and edit transactions directly in a register using the buttons provided at the bottom of the register window. You can also scroll up or down in the register to find a particular transaction, then view or edit it in the corresponding form by clicking on the Edit button.

By the way, registers are not limited to bank accounts. *Every* asset, liability, and equity account has its own register (income and expense accounts do not).

How to Open or Switch to a Form

Here are the various ways you can open a QuickBooks form:

- **From the Activities menu.** All forms are represented by one of the Activities menu choices. For example, to open the Checks form you can choose Activities|Write Checks.

- **With a hotkey combination.** Some forms have a hotkey assigned to them. For example, you can open the Checks form by typing Ctrl-W (press the "W" key while holding down the Ctrl or Control key).

- **From a register.** When the cursor is positioned on a transaction in a register, you can open the form associated with the transaction by either (1) double-clicking the transaction, (2) typing Ctrl-E, (press the "E" key while holding down the Ctrl or Control key), (3) clicking on the Edit button at the bottom of the register window, or (4) choosing Edit|Edit *formname*, where *formname* names the appropriate form.

How to Open or Switch to a Register

- **From the Chart of Accounts window.** With the Chart of Accounts window open, click on any asset, liability, or equity account so that it is highlighted. You can then open a register window for that account by either (1) double-clicking the account name, (2) typing Ctrl-R (press the "R" key while holding down the Ctrl or Control key), (3) clicking on the Register button in the icon bar, or (4) choosing Activities|Use Register.

- **From a form.** When a form is displayed, its corresponding Register window can be opened by either (1) typing Ctrl-R (press the "R" key while holding down the Ctrl or Control key), (2) clicking on the Register button in the icon bar, or (3) choosing Activities|Use Register.

CHAPTER 5

Farm Expenses & Payables

This chapter introduces QuickBooks forms and techniques for paying farm business expenses, including checks, bills, and credit card charges. You also write checks and pay bills for things other than farm expenses of course, such as loan payments or to purchase farm assets, but those are dealt with in later chapters.

Where to Enter Farm Expenses ...as Checks? ...Bills? ...Credit Card Charges?

Problem — *I could enter farm expenses on several different QuickBooks forms: Checks, Bills, Credit Card Charges, or Purchase Orders. When should I use each of these forms, and why?*

Solution — The form you should use depends on the payment method, when payment will be made (now vs. later), and several other factors described in this section.

Discussion — Most farm business expenses are entered on either the Checks form or Bills form. A growing number of farm expenses are paid by credit card and those are entered on the Credit Card Charges form. Because most farm purchases do not involve QuickBooks inventory items, the Purchase Orders form is rarely used.

When to Use the Checks Form

Use the Checks form when you pay an expense directly by check, regardless of whether you have hand-written the check or you will print the check from QuickBooks. QuickBooks' Checks form is laid out just like a real-world check, so entering checks is familiar and easy.

Warning: do not *directly* enter checks to pay bills that have been entered on the Bills form. For those you must generate checks by using the Pay Bills activity

93

(Activities|Pay Bills), so QuickBooks can associate the payment with the bill and mark the bill as paid.

When to Use the Bills Form

Use the Bills form to enter purchases for which payment will be made at a later date.

Most farm expense items are bought on credit terms. That is, you pay for the items some time after receiving them, usually after receiving a statement of account from the vendor (dealer, supplier). Entering bills in QuickBooks gives you an easy, organized way to keep track of how much is owed to which vendors, and when payment needs to be made.

When to Use the Credit Card Charges Form

Use the Credit Card Charges form to enter farm business purchases made by credit card.

You can enter charges and credits during the month, or wait until the credit card statement comes, or both. After entering all of the months charges and credits, let QuickBooks help you reconcile the credit card account—it works just like the process of reconciling a checking account, which is described on page 97.

When to Use the Purchase Orders Form

Most farm businesses use the Purchase Orders form rarely or not at all—you won't miss any big benefits if you avoid it altogether.

Purchase orders are mostly meant to work with QuickBooks' inventory system, to allow keeping track of items on order, and are not a necessary part of accounting with QuickBooks. However, they do give you a way to keep track of the various things you may have ordered or booked for purchase, such as seed. (See page 123 for an example.) They're also useful if part of your business involves purchasing items to sell at retail.

Working with Bank Accounts

Problem
How should I manage my checking account and other bank accounts in QuickBooks?

Solution
Learn to use the QuickBooks features which help keep your transaction entries up to date and "in synch" with your bank account.

Discussion
Checking accounts are the main focus of discussion in this section, but "bank account" means more than just checking accounts: it includes savings and money market accounts, or any other account you might have at a financial institution. Every QuickBooks feature which works with checking accounts—transfers of funds, register windows, account reconciling, etc.—also works with these other types of bank accounts.

Entering checks, deposits, and account transfers, and reconciling a bank account with the bank statement, all work in QuickBooks much like doing the same jobs manually, so you won't need to change many habits to use them. Here's a typical monthly cycle of working with a bank account in QuickBooks:

Enter checks you've handwritten, checks to be printed, **and deposits** (daily or as needed)

Print checks to be printed (if any), and mail to vendors

Receive statement from the bank

Reconcile transaction entries with the bank statement

Setting up Bank Accounts in QuickBooks

Set up a separate QuickBooks account for each bank account belonging to the farm business. Don't include any non-farm accounts, such as a personal checking account.

Most bank accounts are created when you set up a new QuickBooks company, in response to your answers during the interview. But from time to time you may need to add a new account, as when opening a savings account or a new checking account at a different bank.

To set up a bank account:

1. **Open the Chart of Accounts window.**

   Choosing Lists | Chart of Accounts is one way to do this.

2. **Open the New Account dialog.**

   Either type Ctrl-N, or click on the Account button in the lower part of the Chart of Accounts window and then select New from the pop-up menu.

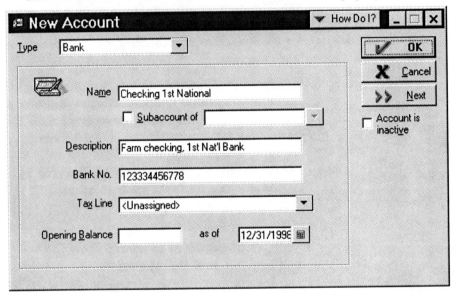

3. **Fill in the fields as desired.**

   Here are comments on some of the important fields:

   Type
   Choose Bank. (Note: the Bank account type is a *current asset* account type.)

   Name
   Use an account name that uniquely identifies the account, such as "Checking/First National" or "Farm Checking", especially if you have more than one bank account.

   Subaccount
   Usually a bank account should not be a subaccount of another account. An

exception would be using one main (parent) account to summarize the balance of several individual bank accounts. Each bank account would be made a subaccount of the main account. The main account's balance (which is always displayed in the Chart of Accounts window) would represent the combined balances of the individual bank accounts.

Bank No.

Nothing is required in this field, but you may enter your bank account number here for quick reference.

Tax line

Do *not* choose a tax line—they are normally only assigned to income or expense accounts.

Opening Balance

This should *always* be zero for a newly opened bank account—your first deposit or transfer of funds into the account will give it an initial balance. Only enter an opening balance in this field if you are setting up a bank account that already existed prior to the company's QuickBooks start date.

4. **Click OK to save the new account.**

Reconciling a Bank Account

Reconciling bank accounts should be part of your regular cycle of accounting activities because it verifies the accuracy of your transactions, and the bank's. It's best to reconcile soon after receiving the bank's statement of account, while details of recent transactions are still fresh in your mind—in case there's a problem you need to track down. Reconciling takes very little time if your transaction entries are up to date in QuickBooks. If they're not, and you have a lot of transactions to enter, reconciling may get postponed for longer than it should be.

Here are the basic steps for reconciling a bank account:

1. **Choose Activities|Reconcile.**

   The Reconcile window will open.

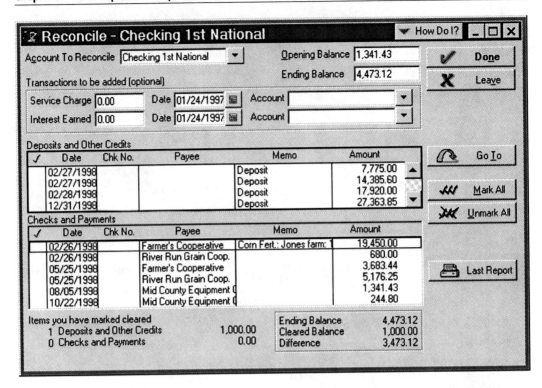

2. In the Account to Reconcile field, select the bank account you want to reconcile.

3. Compare the amount in the Opening Balance field with the opening balance shown on your bank statement.

QuickBooks uses the Ending Balance from the prior reconciliation as the Opening Balance for the current one, so the amount in the Opening Balance field should match the opening balance on the bank statement. However, if you've just created a new company file or if this is the first time for reconciling a new bank account, there may be reasons why the two don't match. Consult QuickBooks Help for ideas on what to do.

4. Enter the ending balance shown on your bank statement in the Ending Balance field.

5. Enter bank service charges and/or interest earned, if any, in the "Transactions to be added" area, and select the accounts to which you want them posted.

You can use the Service Charge and Interest Earned fields to enter *any* expenses or income associated with the bank account. Example: if your bank applies two different types of service charges that you want to keep track of separately, you could enter one of them in the

Service Charge field and the other in the Interest Earned field. (Either or both fields may be used for entering income or expense.)

Some users prefer to enter service charges as checks and income as a deposit (in the Write Checks and Make Deposits windows) so the bank name can be entered in the payee field.

6. Mark all of the "cleared" transactions—the ones that appear on the bank statement.

   Scroll through the Deposits and Other Credits list, and the Checks and Payments list, clicking on transactions that match those listed on the bank statement. A ✔ appears to the left of transactions you've marked as cleared.

   If some transaction amounts don't match, correct the transactions. If you find transactions on the bank statement that have not yet been entered in QuickBooks, open the Write Checks or Make Deposits form and enter them. Then return to the Reconcile window. It is automatically updated to include the new transactions, which you may then mark as cleared.

An easy way to view or edit a transaction listed in the Reconcile window is to double-click it, or highlight the transaction and then click on the Go To button.

7. When you have finished marking "cleared" transactions, look at the Difference amount displayed near the bottom of the Reconcile window.

   • **If the Difference is 0.00,** you have successfully reconciled the account with the bank statement. Click on the Done button.

   • **If the Difference is *not* 0.00,** the account does not balance for the period of time covered by the statement. You need to either (1) find and fix the problem such as a missing transaction, an incorrect transaction amount, or a bank error, or (2) adjust the account balance by the amount of the difference if it's not large enough to justify spending time to find it.

QuickBooks can automatically adjust the bank account balance.

When you click on the Done button in the Reconcile window, if the Difference amount is not zero QuickBooks will ask if you want the account balance adjusted automatically. Click OK to have QuickBooks make the adjustment.

> QuickBooks uses the Opening Bal Equity account as the offsetting account for automatic bank account adjustments.

Entering Checks

Problem

Use the Checks form to enter checks—both for checks QuickBooks will print and for checks you've written by hand.

Solution

The Checks form is formatted to look like a check. This makes entering checks in QuickBooks as familiar as filling in the blanks on a handwritten check. The Checks form also gives you a place to enter the quantities of things you have purchased if you wish.

Discussion

QuickBooks lets you enter checks you've written by hand, and checks that will be printed on your printer. Selecting the "To be printed" check box on the Checks form marks a check for later printing. (When entering a check you've handwritten, be sure the "To be printed" box is not checked.)

When to Enter Checks

Checks you will print from QuickBooks must be entered before you can print them. That's actually one of the benefits of printing checks from QuickBooks: it automatically promotes keeping your records up to date.

Handwritten checks should be entered soon after you've written them. How soon is up to you, but consider this: many people discover they no longer need to keep track of their bank balance elsewhere, once they've gotten into a habit of keeping their checkbook entries up to date in QuickBooks. QuickBooks becomes the master copy of their check register, and isn't subject to the usual math errors and other mistakes inherent in keeping a bank balance by hand.

When **NOT** to Enter a Check!

Do not enter a check for anything you've entered as a bill (i.e., in the Enter Bills window). If you do, QuickBooks won't be able to match your check with the bill and will not show that the bill has been paid—the bill will remain in Accounts Payable indefinitely unless you delete it. Bills should only be paid using the Activities|Pay bills menu selection, which will generate the entries for checks or other types of payment.

How to Enter a Check

1. Choose Activities|Write Checks

The Write Checks window will open.

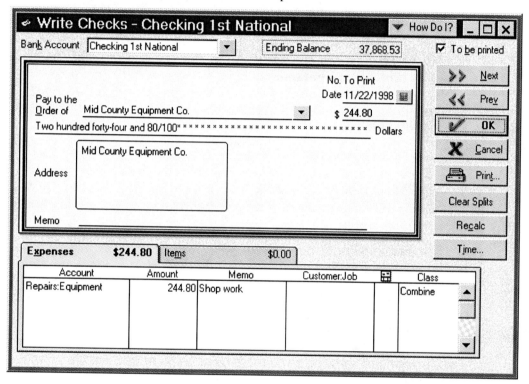

2. Fill in the form's fields.

Here are comments about some important fields:

Bank account

If you have more than one farm bank account, be sure to select the appropriate account here. Otherwise the check may be posted to the wrong account.

If you only use one bank account...

You may want to turn on the QuickBooks preference which begins check entry with the cursor placed in the payee ("Pay to the Order of") field, bypassing the Bank Account field. Here's how: choose File|Preferences, and in the Preferences window select the Checking icon in the scrollable box. Click on the Company Preferences tab (in QuickBooks 6 and later) and click on the "Start with payee field on check" preference item.

If you use *more than one* bank account...

Color-coding your bank accounts can reduce errors. Select a particular bank account and open the Write Checks window or Register window, then choose Edits|Change Account Color and select the color you want to associate with the account. Thereafter, the background color of forms and registers will be set to this color when you are working with that bank account.

You may color-code *any* balance sheet account (assets, liabilities, equity accounts). So this same technique works for credit card accounts, loan accounts, etc., providing visual confirmation that you're working with the right account.

No. (check number)
If you are recording a handwritten check, enter the check number here. If QuickBooks will print the check there's no need to enter a number: Quick-Books will automatically enter the proper check number in this field when the check is printed.

To be printed
Place a check mark in this field to mark the check for later printing by Quick-Books. When this field is checked, the No. (check number) field is disabled.

Pay to the Order of (payee)
This is the person or business to whom you're writing the check. Though QuickBooks permits leaving this field blank, don't! Get into the habit of entering names whenever requested in QuickBooks, to keep the detail in your transactions.

Fields on the Expenses Tab and Items Tab

If QuickBooks' inventory features are turned on, an Expenses tab and an Items tab will appear on the Checks form. (In recent QuickBooks versions both tabs are always present.)

You can enter expenses on either or both tabs—their combined total must equal the total for the check. The Expenses tab is for recording expenses without quantity information. The Items tab lets you record purchases of specific items and include the quantity purchased. *Items* are names you set up in QuickBooks to identify specific things you buy and sell. (See Setting Up and Using QuickBooks Items on page 68.)

Expenses tab fields:

Account

Usually this will be an expense account, such as Seed Expense, though sometimes you will choose other account types—as when purchasing assets or when writing a check to transfer funds from one bank account to another.

Class

You may select a class here, which allows gathering enterprise information and other special information from your transactions, as discussed in chapter 3. If you want to use classes but don't see a Class field, you need to turn on class tracking in the Preferences window (File|Preferences).

Items tab fields:

Item

Select the appropriate item name from the drop-down list, or type a new name and QuickBooks will ask if you want to set up a new item definition.

Qty (quantity)

This is where you enter the purchase quantity—the number of bushels, pounds, tons, or head bought.

Be sure to use the *same unit of measure* when entering quantities for an item. Otherwise, the quantity information won't be very useful. For instance, if you record some feeder calf purchases in pounds and other feeder calf purchases in hundredweights, the resulting "pounds and hundredweights" total will be meaningless.

Cost

You may enter the per-unit cost of the item, such as the per-gallon cost of diesel fuel, in this field. When you first select an item for the transaction this field may be filled with a cost figure from the item's setup information. If that cost figure isn't appropriate for the current transaction, override it by typing in a new cost. Or, you only enter values in the Qty (quantity) and Amount fields, QuickBooks will automatically calculate and fill the Cost field for you.

Amount

This field holds the dollar amount for the transaction line. You may enter values in the Qty and Cost fields and let QuickBooks calculate the Amount for you, or you can type in an Amount.

How the Qty, Cost, and Amount fields are related

In any Checks or Bills form you only need to enter information in two of these fields—or just one field if it's the Amount field. Here are the possible combinations:

Qty and Cost - Fill these fields and QuickBooks will calculate and fill in the Amount field.

Qty and Amount - Fill these fields, and QuickBooks will calculate and fill in the Cost field.

Amount - QuickBooks will accept a transaction with just the Amount field filled if that's how you want to enter it.

These three fields are always linked in this way, so be careful when you edit transactions. For example, suppose you originally enter the Qty and Amount on a transaction line and QuickBooks calculates and fills in the Cost field. If you later go back to that transaction line and change the number in the Cost field, the Amount will be recalculated automatically...so be sure that's what you want to do.

Class
You may select a class here, which allows gathering enterprise information and other special information from your transactions, as discussed in chapter 3. If you want to use classes but don't see a Class field, you need to turn on class tracking in the Preferences window (File | Preferences).

3. Click Next or OK to save your check entry.

Deducting Cash Discounts from a Check or a Bill

Problem

When the farm supply dealer gives me a cash discount, how do I enter that on a check or bill?

Solution

Enter the gross expense amounts on the check or bill as usual, then add a separate line for the discount and enter it as a negative amount.

Discussion

QuickBooks allows entering lines with negative dollar amounts in checks and bills, so long as the total amount is positive.

The following examples show how to enter a cash discount on a check. Cash discounts are entered on bills in exactly the same way.

How to Enter a Cash Discount on a Check or Bill

Entering a discount on the Expenses tab

1. Set up an income account in your Chart of Accounts for recording discounts, if you don't already have one.

   Name the account something like Cash Discounts, or Discounts & Rebates.

2. Click on the Expenses tab of the check or bill, to select it.

3. Use the Cash Discount account on any detail line in a check or bill to record a cash discount.

   Be sure to enter the discount amount as a *negative* number. This will credit the cash discount account with income and deduct the discount from the form's total, as shown here:

Expenses	$1,680.70	Items		$0.00	
Account	Amount	Memo	Customer:Job		Class
Seeds & Plants	960.00	10 seed corn 6303			Corn
Seeds & Plants	755.00	10 seed corn 7330			Corn
Cash Discounts	-34.30	Cash discount, seed			Corn

On forms where QuickBooks provides a **Recalc button,** you can click on it to recalculate the form's total.

Entering a discount on the Items tab

4. Set up an income account in your Chart of Accounts for recording discounts, if you don't already have one.

   Name the account something like Cash Discounts or Discounts & Rebates.

5. Add an item to the Items List for entering discounts, if you don't already have one

   Select Other Charge or Non-inventory Part as the item type. Name the item something like Discount, and select your cash discounts income account in the Account field.

QuickBooks has a Discount item type, but *do not* assign it to cash discount items you set up to use on checks and bills. The Discount item type can be used on invoices but *QuickBooks does not allow using it on checks or bills.*

6. Click on the Items tab of the check or bill, to select it.

7. Add a line to the transaction for the discount by selecting your discount item and supplying a <u>negative</u> dollar amount.

As shown here, the discount subtracts from the Items tab's total:

Expenses	$0.00	**Items**		$1,680.70		
Item	Description	Qty	Cost	Amount	Customer:Job	Class
Seed corn	Seed corn 6303	10	96.00	960.00		Corn
Seed corn	Seed corn 7330	10	75.50	755.00		Corn
Cash discnt	Cash discount		-34.30	-34.30		Corn
Select PO				Receive All		Show PO

Printing Checks from QuickBooks

How should I use QuickBooks for printing checks?

You can either print checks one at a time, or in a batch.

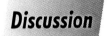
Printing checks one at a time is done by clicking on the print button in the Write Checks window. Printing a batch of checks involves two steps. First, mark each of the checks you want to print by selecting the "To be printed" check box in the Write Checks window. Then choose the QuickBooks menu command for printing the batch of marked checks.

Check forms preprinted with your farm business name, sequential check number, and magnetic-ink bank and account numbers are available from several sources (see Appendix A).

How to Print Checks One at a Time

To print a check directly from the Write Checks window:

1. **Load a check form into your printer.**

   You may load more than one form if you plan to print several checks.

2. **Click on the Print button in the Write Checks window.**

   QuickBooks will display a Print Check window asking for the check number.

3. **Enter the correct check number, then click OK.**

   The number you enter should match the number on the check form loaded in your printer. If the check form is unnumbered, just enter the next check number you want to use.

   QuickBooks will display a Print Checks dialog.

4. **Click on the Print button in the Print Checks dialog to print the check.**

5. **Don't forget to remove unused check forms from your printer if you plan to use it for other types of printing.**

How to Print Checks in a Batch

The first requirement of printing checks in a batch is to have marked the checks for printing. Do this by selecting the "To be printed" check box in the Write Checks window as you enter each check. Then print the entire batch of marked checks, like this:

1. **Load check forms into your printer.**

   Load plenty of forms—you can remove the unused ones when you're done printing checks.

2. **Choose File|Print Forms|Print Checks.**

   QuickBooks displays the Select Checks to Print window, where you can select some or all of the marked checks for printing.

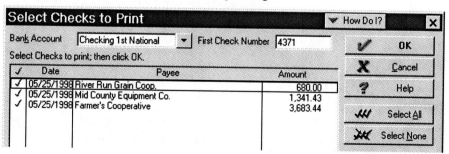

3. Select the appropriate bank account, beginning check number, and the checks you want to print.

Bank account

This field lets you choose the bank account for which you want to print checks.

First check number

Enter a check number matching the number of the first check loaded in your printer. If the check form is unnumbered, just enter the next check number you want to use.

Select checks to print

You can select any or all of the checks by clicking on them individually, so that a ✔ appears in the leftmost column. The Select All and Select None buttons give you a quick way to select or deselect all checks.

4. Click OK.

   QuickBooks will display the Print Checks dialog.

5. Click on the Print button in the Print Checks dialog to print the checks.

6. Don't forget to remove any unprinted check forms from your printer if you plan to use it for other types of printing.

 Consider using windowed envelopes for mailing checks. Their cost is soon repaid by eliminating the need to address envelopes by hand.

Reprinting Checks

To reprint one check, find the check in the Write Checks window, then click on the Print button. To reprint a batch of checks, find them one by one in the Write Checks window and re-select their "To be printed" check boxes, then print the batch as described above.

If you print to a different check number—such as when a check was accidentally damaged and you are printing a new one—QuickBooks will update the check entry's No. (number) field with the new check number, based on the starting check number you supply.

Working with Bills

 Bills is the QuickBooks term for what accountants more typically call accounts payable—amounts you owe for purchases on credit.

Problem

How do I use QuickBooks to keep track of my bills, remind me when bill payments are due, and forecast my cash requirements for bill payments?

Solution

Enter bills soon after you receive them, so QuickBooks can list them in the Reminders window. Use the Reminders window to stay aware of the due dates and payment amounts of bills coming due over the next few weeks. When the due date arrives, use Activities|Pay bills to automatically prepare checks for the bills you want to pay.

Discussion

QuickBooks' features for managing bills are one of the biggest benefits of using QuickBooks in a typical farming or ranching operation. Why? Because managing and paying bills is the bookkeeping activity that typically demands more time and effort than any other. You may sell grain or livestock once a month or even less often. But hardly a week passes without the need to pay bills.

Even if you are a cash basis record keeper you should use QuickBooks' bill management features. (Just be sure to select the Cash Basis option when preparing reports, which causes accrual entries—bills you have entered but not yet paid—to be ignored for the report.) The biggest hurdle may be the need to change some of your current habits for how you go about paying bills. But the changes will be well worth the effort.

Here's a typical cycle for managing bills in QuickBooks:

Make purchases on credit from vendor

Receive invoices and/or statement from vendor

Enter invoices and credits on the Bills form

Watch the Reminders window for bill due dates
Print a Cash Flow Forecast (optional) to see how bill payments may affect your bank balance

Prepare payments using the Pay Bills window

Send payment to the vendor

Accounting 101: The accounting effects of checks vs. bills

Whether you enter a check or a bill, the detail lines of the entry are posted to expense accounts in exactly the same way. But the *offsetting amount* is posted differently for checks and for bills. For a check it is posted as a withdrawal from the checking account, and for a bill it is posted as an increase in Accounts Payable, the amount you owe others.

How to Manage Bills in QuickBooks

The following points outline one approach for managing bills in QuickBooks. It's not the only approach but provides a good starting point for deciding how you want to handle bills in QuickBooks.

1. **Enter bills within a few days of receiving them.**

   - This assures bills get listed in the Reminders window, described below, so you'll be less likely to miss bill due dates.

   - It lets you enter the bill while details of the transactions are fresh in your mind. You'll do a better job of accounting and catching billing errors then, than if you put off the job for a few weeks.

   - Make entering bills well ahead of their due date a bookkeeping *priority*. Entering bills ahead of time is more likely to result in good records than entering bills in a hurry on the day payment is due. It's a good "rainy day" job—something best done when you can spend time on the important details.

   - Use QuickBooks' Automatic recall and Memorized Transactions features, discussed in chapter 9, to make entry of regularly-recurring bills quick and easy.

   - With the necessary expense details entered in advance, being short of time on the bill's due date won't be a problem. All that will remain to do then is print and mail a check.

2. **Assign an appropriate due date as you enter each bill.**

   QuickBooks displays bills in the Reminders window *sorted according to their due dates*. So if you fail to enter a correct due date you may miss paying the bill on time.

Assign due dates that are a few days ahead of the actual due date, to allow yourself time for getting the payment prepared and for the payment to travel through the mail.

Unless you are using QuickBooks daily, **group payment due dates** so you only need to pay bills once or twice a week.

3. **Make sure the Reminders window is set up properly.**

   Here are some recommended preference settings for the Reminders window:

   • Have QuickBooks always open the Reminders window each time you open the company file (page 28).

   • Choose the Show List option for Bills to Pay, so bills will be listed individually in the Reminders window (page 28).

   • Allow an adequate number of days to be reminded ahead of bill due dates (page 28). The less frequently you use QuickBooks, the larger this number should be. Most users prefer to be reminded of due bills 15 to 30 days in advance.

4. **Use QuickBooks frequently enough that the Reminders window keeps you aware of due dates.**

   Using QuickBooks at least once a week should keep you aware of important payment dates.

If you don't plan to use QuickBooks often, you may print out the Reminders list and post a copy where you will see it daily—beside a wall calendar, on the refrigerator, etc.

To print the list, click on the Reminders window to make it the topmost window (in front of all the others), then choose File|Print list.

5. **If you want a better forecast of cash requirements, print a copy of the Cash Flow Forecast report.**

   The Cash Flow Forecast report (Reports|Other Reports|Cash Flow Forecast) can show cash requirements over any period of time you select.

6. **Prepare bill payments in the Pay Bills window.**

   Selecting Activities|Pay Bills opens the Pay Bills window, where you can select which bills to pay. QuickBooks then makes check entries (or other forms of payment, such as credit cards) to pay them, and marks the bills paid.

Anything you have entered as a bill should *only* be paid using the Pay Bills window. Otherwise QuickBooks won't be able to connect the payment with the bill, and it will remain in the Reminders window as an unpaid bill.

7. **Send payment to the vendor(s).**

This step will depend on the payment method you chose in the Pay Bills window:

- **Checks to be printed by QuickBooks** - Print out the checks generated by the Pay Bills activity, then send them to the vendor(s).

- **Checks to write by hand** - Write checks from your checkbook to match the checks generated by the Pay Bills activity. You can either write checks in the same order as the generated checks or edit the check numbers of the generated checks to match those of the checks you write.

- **Credit card payments** - If you chose to pay by credit card, use whatever means are necessary to get the credit card payment information to the vendor(s). (The Pay Bills activity already created payment entries in your credit card account.)

- **Online payment** - The payment information will be transferred the next time you go online, and payments will be sent to the vendor(s).

Entering Bills from Vendors (Accounts Payable)

Problem

How and when should I enter bills in QuickBooks? Should I enter them throughout the month, or wait until I receive the dealer's itemized statement at the end of the month? Should I enter each invoice separately or can I enter all of the transactions from the dealer's statement as a single bill?

Solution

Entering bills directly from the dealer's monthly statement is fine, if it's an itemized statement containing sufficient detail. Entering a separate bill to match up with each of the dealer's invoices is a good idea, though not absolutely necessary. Having one bill for each of the dealer's invoices may make it easier to catch billing errors.

Discussion

An itemized monthly statement may be the first transaction detail you see for the month's purchases from a particular dealer. Even if you pick up a copy of the invoice or ticket for each purchase, you probably wait until you receive the dealer's monthly statement before entering any of them—that's the most common approach.

Should you enter a separate bill for each invoice, or enter the entire month's activity as one bill? As you might guess, the answer is "It depends..."

Entering a separate bill for each invoice is the more correct approach, from an accounting standpoint. You may enter a lot of individual bills that way, but that's also an advantage: each bill you enter will have fewer detail lines, making it easier to catch errors—yours *or* the dealer's. If anything has been entered incorrectly the bill's total won't match the dealer's invoice.

Another reason to enter a separate bill for each invoice is when the dealer's credit terms are offered invoice-by-invoice. If the terms are "net due in 30 days", then having a separate bill for each invoice lets you postpone paying individual invoices until they are actually due.

On the other hand, entering one bill for the entire month's purchases from a vendor may seem like the easy way to go. You just keep entering transaction lines from the statement until you're done, without the interruption of having to frequently move on to a new bill. But this approach won't give you a separate total for each bill—nothing to compare against the individual invoice totals on the dealer's statement. If there is an error in your "everything-as-one-entry" bill and it contains a lot of detailed transaction, don't expect the error to be easy to find.

Basics of Entering Bills

1. Choose Activities | Enter Bills

QuickBooks opens the Enter Bills window.

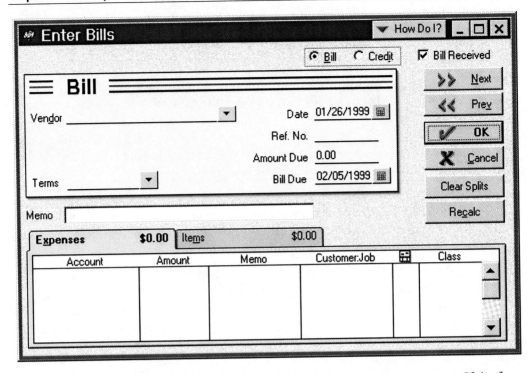

This window serves for entering either bills or credits on account. If the form title reads "Credit", select the Bill option at the top to change the title to "Bill":

2. Fill in the fields as desired.

Entering a bill is a lot like entering a check—the two forms are almost identical. The important difference is that you need to enter the due date for a bill, or QuickBooks won't be able to provide a note in the Reminders window when the bill comes due.

Here are notes about important fields on the Bills form.

Vendor
Because a bill records your liability to a particular vendor, be sure you've selected the vendor's name here.

Date
This should normally be the date the invoice was issued by the vendor. The Bill Due date is calculated based on this field and the Terms field.

Terms
Select the vendor's credit terms here—such as "net 30", for amounts due 30

days from the invoice date. The Bill Due date is calculated based on this field and the Date field.

Bill Due

This field reflects the date when the bill is due. It is calculated from the values you've entered in the Date and Terms fields, but you can override the calculated date by typing in a different one.

This is the date QuickBooks will display for the bill in the Reminders window. You may often find it necessary to override the calculated date, to enter a date that's early enough to assure the payment reaches the vendor on time.

Expenses tab fields & Items tab fields

The lower part of the Bills form is where you enter individual transaction lines. It works just like the lower part of Checks form. For detailed information about fields on the Expenses tab or the Items tab see Entering Checks earlier in this chapter.

3. Click OK or Next to save your bill entry.

Entering & Applying Credits on Account (Issued by a Vendor)

Problem

I returned seed to the dealer and they issued me a credit on account. How do I record the credit? How do I apply it to other bills at the dealer?

Solution

Enter credits on the Bills form (it lets you switch between entering either bills or credits). Then use the Pay Bills activity to apply the credits to unpaid bills.

Discussion

The Bills form has an option box at the top that lets you switch between entering bills and entering credits. Use the credits option to enter credits on account issued by a vendor.

How to Enter a Credit on Account

1. Select Activities|Enter Bills.

The Enter Bills window will open.

2. Click on the Credit option in the upper part of the Enter Bills window:

The form title changes from "Bill" to "Credit", to indicate that you're entering a credit.

3. Fill in the form with the details of the credit.

You use the same accounts and/or items (in the lower section of the form) to enter the details of the credit as you would for a bill. The result, however, is exactly the opposite. Instead of recording expenses, a credit reverses them.

When a credit is for returned merchandise, select the account (on the Expenses tab) or the item (on the Items tab) matching what you returned. In this example a credit is applied to the Seed Expense account, because it was a issued for returned seed corn:

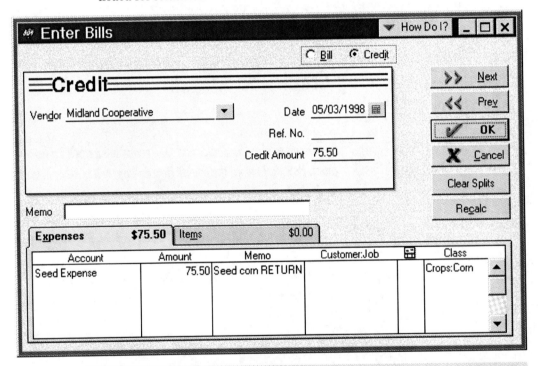

 Typing the Memo for returned items in ALL CAPITALS is a way to make them MORE NOTICEABLE on reports.

On a report showing transactions in the Seed Expense account, you can see that the credit for returned seed subtracts from the account total, as it should:

Seeds & Plants					
Check	03/15/1998	Midland Cooperative	Seed corn 6303	Checking 1st National	960.00
Check	03/15/1998	Midland Cooperative	Seed corn 7330	Checking 1st National	755.00
Credit	05/03/1998	Midland Cooperative	Seed corn RETURN	Accounts Payable	-75.50
	Total Seeds & Plants				1,639.50

When a credit is issued for something other than returned merchandise, a bit more little thought may be necessary to decide where to post it.

Suppose for example, that you win a door prize at the local equipment dealer's winter meeting: $100 worth of shop labor. The dealer gives the prize as a $100 credit on your account. When you enter the credit in Quick-Books you might assign it to an income account—Cash Discounts & Rebates Income would be a logical choice. Or, you might consider the door prize a reduction in repair expenses and assign it to Repairs Expense, which would subtract $100 from the total for your Repairs account, similar to the Seed Expense example above.

Applying a Credit to Outstanding Bills

Just *entering* a credit does not automatically *apply it* to any unpaid bills. For that you need to use the Pay Bills window (Activities|Pay Bills) as described on page 120.

The best time to apply a credit is when it's time to make a payment. The reason is that QuickBooks automatically generates a payment entry (i.e., a check) if the amount of the credit is less than the total amount due on the bills you apply it to.

Handling a Refund of Your Account's Credit Balance (from a Vendor)

Problem
I had a credit on account at the local farm supply dealer, but then they issued a check to refund the credit balance to me. How to I record the refund?

Solution
Include the refund check in a deposit entry, just like any other check. The details you need to enter will depend on whether you previously entered the credit balance in Quick-Books.

Discussion
If the credit balance *was not* entered in your QuickBooks records, simply enter the refund as income or as a reduction in some expense account. But if the credit balance *was* entered in QuickBooks, the deposit entry needs to offset or "cancel out" that credit balance. (If not, your records will count the refund as income twice!)

Accounting 101: What's happening when a vendor refunds your account's credit balance...

The refund check represents a conversion of one asset type to another. Your credit balance on the vendor's account was an asset. (In Quick-Books it's actually recorded as a negative liability, but the result is the same.) By issuing a refund the vendor was converting the credit balance to another type of asset: cash which could be deposited in the farm bank account.

If the Credit was not Entered in QuickBooks...

Because no credit has been entered in QuickBooks, there's no credit entry to reverse when you record the refund (i.e., you don't need to use the Pay Bills window). Just include the refund check in any deposit and assign the account which best represents what the refund was for—Seed Expense, Cash Discount Income, etc.

If the Credit was Entered in QuickBooks...

Because you've entered a credit in QuickBooks, *you have already recorded the income* (or decrease in expense) that was the reason for the credit or refund. So when you deposit the refund check, the deposit entry should only offset the Accounts Payable balance created by entering the credit. The deposit *should not* record the income again, because then the income would be entered twice. Here are the steps involved:

1. **Deposit the refund check, selecting Accounts Payable as the account in the From Account field.**

   Be sure to choose the appropriate vendor name in the Received From field. Otherwise, QuickBooks won't match up the refund with the correct vendor account, and the refund check will not appear in the Pay Bills window as described in the next step.

   Here's how the detail area of the Deposits window might look after entering the deposit:

Received From	From Account	Memo	Chk No.	Pmt Meth.	Class	Amount
Midland Cooper	Accounts Payat	Refund acct bal		Check		75.50

	Deposit Subtotal	75.50

2. **Match the credit balance with the refund check, in the Pay Bills window.**

   Open the Pay Bills window and select both the refund check and the line representing the credit balance you entered earlier—as if both were bills to be paid—then click OK. The refund will offset or "cancel out" the credit balance, and both will disappear from the list. (If you get a warning about using a credit memo to pay a bill, ignore it.) You'll find details on using the Pay Bills window in the next section.

Paying Bills and Applying Credits on Account

Problem	*Once I've entered a bill in QuickBooks, how should I pay it? How can I apply a credit on my account to outstanding bills?*

Solution	Use the Pay Bills window to apply credits to outstanding bills and to prepare bill payments.

Discussion	When you send payment to a vendor QuickBooks must connect it with the specific bill you are paying, and the Pay Bills window is where this happens. By allowing you to select the bills to pay and the credits to apply, and then generating the payments (checks, etc.) for you, QuickBooks assures that (1) your payments will be for the correct amounts, (2) the appropriate bills are marked as paid, and (3) the payment amounts are removed from your Accounts Payable balance for the vendor.

Because these jobs are essential, *bills you have entered in QuickBooks should only be paid from the Pay Bills window*—not by any other means, such as directly entering a check. Payable amounts you *have not* entered as bills in QuickBooks can be paid directly by check, of course.

 If you have accidentally entered payment for a bill without using the Pay Bills window, see the discussion on page 123.

How to Apply Credits and Pay Bills

Bills and credits you've entered appear in the Pay Bills window. There you can select the bills you want to pay and the credits you want to apply to them. QuickBooks applies the credits to the bills, then generates checks or other payment entries for the remaining unpaid amount.

Here are the steps:

1. **Select Activities|Pay Bills.**

   The Pay Bills window will open. Here's an example, showing a $75.50 credit and a bill, both for Midland Cooperative.

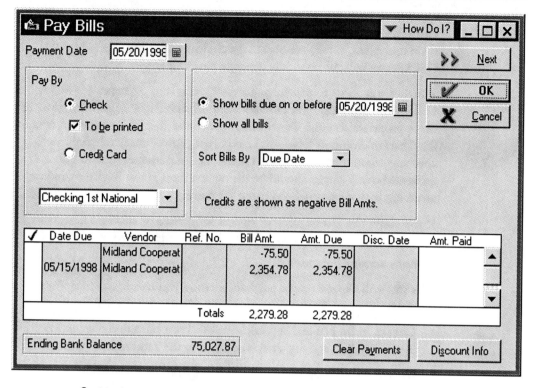

2. Under Pay By, choose the payment type, the account to use, and whether or not to print checks (if the payment type is Check).

This step is necessary because QuickBooks will automatically generate payment entries (checks or credit card entries) for the unpaid balance of bills you select.

Note that you can choose to have QuickBooks generate check entries but not print them—use this option if you will handwrite the checks.

3. In the ✔ column, click on the bills you want to pay and the credits you want to apply.

Grouping bills/credits by Vendor

Normally the Pay Bills window lists all bills first, then all credits. Sometimes it's helpful to see the bills and credits for a vendor together, which you can do by selecting Vendor in the Sort Bills By field.

Also, to make partial payment on a bill or to apply only part of a credit, enter a dollar amount in the Amt. Paid column.

4. Click OK to process your selections and close the Pay Bills window.

QuickBooks will apply the selected credits, if any, to the selected bills. Where credits are more than the bill amounts for a vendor, part of the credit will remain unapplied. Where credits are insufficient to pay the selected bills, QuickBooks will automatically generate payment entries.

Whether you chose to generate checks or credit card payments (or online banking transactions), you can view them by opening a Register window for the appropriate account. To see newly generated checks, for example, open the Chart of Accounts window (Lists|Chart of Accounts), highlight the checking account by clicking on it, then choose Activities|Use Register. The generated check entries should be the newest ones in the Register window, assuming they were generated using the current date.

If you chose the "To be printed" option for generated checks, a Checks to Print line reminder will also appear in the Reminders window.

5. **Prepare vendor payment(s).**

   This step will depend on the payment method you chose in the Pay Bills window:

   - **Checks to be printed by QuickBooks** - Print out the checks generated by the Pay Bills activity, then send them to the vendor(s).

   - **Checks to write by hand** - Write checks from your checkbook to match the checks generated by the Pay Bills activity. You can either write checks in the same order as the generated checks, or edit the check numbers of the generated checks to match those of the checks you write.

   - **Credit card payments** - If you chose to pay by credit card, use whatever means are necessary to communicate the credit card payment information to the vendor(s). (The Pay Bills activity already created payment entries in your credit card account.)

   - **Online payment** - The payment information will be transferred the next time you go online, and payment will be sent to the vendor(s).

6. **Send payment to the vendor(s).**

Consider using windowed envelopes for mailing checks. Their cost is soon repaid by eliminating the need to address envelopes by hand.

What to Do if You've Paid a Bill Without Using the Pay Bills Window

As mentioned earlier, bills entered in the Enter Bills window should only be paid from the Pay Bills window. But what if you've accidentally paid one of these bills another way, such as by directly entering a check?

If you don't notice the problem until after you've printed or mailed the check, it's easiest to find the original bill in the Enter Bills window or the Accounts Payable register window, click on it to highlight it, then delete it (Edit|Delete Bill). The bill entry is no longer needed if you have entered a check with the same amount of transaction detail.

If you have not yet mailed the check you have two choices: either (1) delete the bill as described above, or (2) void the check and generate a new one from the Pay Bills window.

Using Purchase Orders to Keep Track of Ordered Items

Problem

Can QuickBooks help me keep track of things I've ordered, such as units of seed corn booked in advance of planting season?

Solution

You can use purchase orders for this purpose, however their value for showing total quantities on order is limited, so you'll have to decide if they're worth the effort.

Discussion

A purchase order, or "PO", is a list of things you've ordered from a vendor and agreed-to prices. Traditional business practice is to prepare duplicate purchase orders, mail or fax one copy to the vendor, and keep one in your own files. Used this way a purchase order confirms your order with the vendor and also serves as your own record of what you've ordered.

It's unlikely that you will send many purchase orders to your suppliers—typical supplier relationships in agriculture don't call for that. But using purchase orders in QuickBooks is one way to keep track of things you've ordered or booked, like seed, chemicals, or supplies, plus details like expected delivery dates and notes about each order. A side benefit is that if you later enter a bill to pay for the things listed on a purchase order, QuickBooks can copy the detail lines from the purchase order to the bill, saving lots of typing.

Benefits of using purchase orders in QuickBooks:

◆ Provides an organized way to keep track of items you have on order.

◆ Easy to get a listing of all orders for a specific vendor.

◆ QuickBooks can automatically copy detail lines from a purchase order to a bill when you enter payment for the received items.

Limitations of using purchase orders in QuickBooks:

◆ QuickBooks reports which show quantities on order *only work for Inventory Part type items*.

This limitation is a serious one. Because the QuickBooks inventory system isn't well matched to farm inventories, items you set up for seed, chemical purchases, and the like will typically be Service or Non-inventory Part items—not Inventory Part items. But if you enter purchase orders using those item types, you'll have to view each purchase order individually and manually add up the quantities of items on order. (See page 69 for information about item types.)

A Simple Purchase Order Example: Seed Corn Orders

The following steps illustrate an example of using purchase orders to keep track of the quantity and price of various seed corn hybrids on order from a dealer. Using the Purchase Orders form is similar to working with any other Quick-Books form, like Bills or Checks. However, all of the transaction detail on a purchase order must be entered using items (there are no separate Expenses and Items tabs).

1. **Open the Create Purchase Orders window.**

   Choose Activities | Create Purchase Orders to open the window.

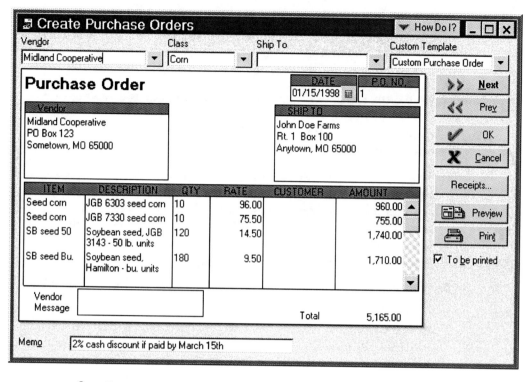

2. **Fill in the form's fields.**

   This example shows entries for booked corn and soybean seed.

   Here are notes about some important fields and other features of the Purchase Order form.

   Vendor

   It is *never* a good idea to enter a purchase order without selecting a vendor, even though QuickBooks allows it.

   Item

   Select items in this column representing the things you have on order.

   If purchase orders are to accurately represent what you have on order, be sure to use items which properly identify them. Note the two detail lines for booked soybean seed in the example. Each line uses a different item—one is for a 50-lb units of soybean seed, and the other is for bushel units.

   Vendor Message and Memo

   If you won't be printing off a copy of the purchase order to send to the vendor, you can use these fields to enter any notes you want to keep, such as the expected delivery date, comments about an early-booking discount, etc..

3. **Click OK to save the purchase order.**

4. **You can find out details about the seed you have on order, at any time.**

If your purchase orders contain Inventory Part type items, an Item Listing report (Reports|List Reports|Item|Item Listing) will show the total quantity on order for each one.

On the other hand, if the items in your purchase orders are Service or Non-Inventory Part items—which will often be the case in a farm business—there's a minor problem: no QuickBooks report can show the total quantity on order for these item types.

To get quantity information for these item types you must look at each purchase order individually, or print them out and manually add up the quantities. A good report for this purpose is the Open Purchase Orders report (Reports|Purchase Reports|Open Purchase Orders). After opening this report you can double-click any purchase order listed in the report window to open it for viewing or printing.

5. **After taking delivery of the seed, enter a bill for it by choosing Activities|Inventory|Receive Items and Enter Bill.**

After you select the vendor's name on the Bills form, QuickBooks will notify you that open purchase orders exist for the vendor and will ask if you want to receive against them. Choose Yes, and a list of the vendor's outstanding purchase orders will be displayed. Select the appropriate purchase order from the list, and QuickBooks will copy its detail lines onto the Bills form, where you can then change or add any details as necessary before saving the bill.

Paying Expenses with the Farm's Cash (Petty Cash)

Problem

We keep some cash on hand to use for miscellaneous farm spending. This cash is farm business cash—used for farm expenses only, and kept separate from our personal cash. How should I enter farm expenses that were paid using this cash?

Solution

Use a petty cash account.

Discussion

Petty cash is a small amount of cash kept on hand for miscellaneous spending. It is also an asset and should be included on the balance sheet. If your farm business keeps its own separate fund of cash on hand, then setting up a petty cash account is the best way to keep track of it and the expenses paid from it.

There's a full discussion of petty cash accounts in Maintaining a Farm Business Petty Cash Fund on page .

Paying Farm Expenses with Non-Farm Funds (Personal Cash, Checks, or Credit Cards)

Problem

If I pay a farm expense with one of my personal checks or with cash out of my own pocket, how can I get the expense entered into the farm's accounts in QuickBooks?

Solution

An easy way to enter farm purchases made with non-farm funds is through a special account created for that purpose, much like a petty cash account. But some situations require other approaches.

Discussion

Your farm accounting records need to include all farm expenses, even those paid from non-farm funds. Otherwise you'll have incomplete expense totals at income tax time and for management purposes. (*Funds* in this case means anything you could use to make a purchase, such as cash, a check, or a credit card.)

The problem is getting expenses paid with non-farm funds entered into the farm business records without incorrectly affecting the balance sheet. How can this be? Since QuickBooks is a double-entry accounting system, every transaction affects the balance of at least two accounts. When you enter a farm expense it is posted to an expense account and to a second or "offsetting" account which identifies the *source* of the funds which paid the expense. Choosing the wrong offsetting account will produce incorrect balance sheet figures.

In QuickBooks, the form you choose for entering a transaction *automatically* determines which offsetting account will be used.. For example, when you enter a check on the Checks form QuickBooks automatically knows to post offsetting amounts to the checking account, as decreases in the checking account balance.

But if this is true, which form should you use to enter expenses paid with non-farm funds? What offsetting account is appropriate? The answer comes from understanding what happens, from an accounting standpoint, when a farm expense is paid with non-farm funds: *it adds to the supply of capital (owner's equity) in the farm business.*

Knowing this gives you two choices for handling the addition to farm business capital:

127

◆ **You can enter the expense as an increase in owner's equity.**

This is what happens most often in sole proprietorships and partnerships. The appropriate offsetting account for an expense paid with the owner's personal funds would be an equity account, to record the increase in owner's equity in the farm business. Actually though, QuickBooks makes the entries for this kind of a transaction easier if you take a slight detour from this plan, and set up a special account to keep track of purchases made with non-farm funds, as explained later in this section.

◆ **You can pay back the capital addition by making a reimbursement for the expense.**

Enter a check from farm checking or other farm business funds to reimburse the person whose funds were used to pay the expense. Corporations must do this, because capital can only be added to a corporation by issuing stock.

The rest of this section describes three ways to enter farm purchases made with non-farm funds:

◆ Entering the purchases in a special "non-farm funds" account

◆ Entering the purchases as capital additions to the farm business

◆ Reimbursing the purchaser from farm checking (or other farm funds)

How you handle farm and personal cash deserves careful consideration. For a recommended approach see The Best Way to Keep Track of Cash Spending and Income, on page 231.

Using a "Non-Farm Funds" Account

Pros:

◆ Easy, because you enter purchases in the Write Checks window.

◆ All purchases made with non-farm funds are entered in a single account, so you always know where to look for information about those transactions.

◆ Allows handling the purchases as either capital additions to the farm business or as reimbursements to the purchaser.

◆ Can be used by sole proprietorships and partnerships.

Cons:

◆ Periodically requires an adjusting entry to record the balance sheet effects of purchases made with non-farm funds.

◆ Should not be used by corporations.

The basics of this approach are that you (1) set up a special-purpose account of the Bank account type, (2) use it for entering farm purchases made with non-farm funds, and (3) periodically make an adjusting entry to transfer the account's balance to an equity account. (The adjusting entry accounts for the additions to farm business capital which result from farm purchases made with non-farm funds.)

This method is a lot like using a petty cash account, described in chapter 9. However, do not use the same account for both purposes—the petty cash account should be separate.

Couldn't I just set up an account for my personal check book, in the farm records?

Yes, if you don't mind having your personal checking account included on the farm balance sheet. But a checking account is not as versatile as a Non-Farm Funds account, because you can't use it to enter transactions involving other types of non-farm funds, like personal cash or credit cards.

You'll find a full discussion about setting up and using a Non-Farm Funds Account in chapter 9, but here's a short example:

1. **Set up a Non-Farm Funds account in QuickBooks, if you don't already have one.**

   Open the Chart of Accounts window, then choose Edit|New account. Select Bank as the account type. Name the account Non-Farm Funds or some other name of your choice, and give it a beginning balance of zero.

2. **When a farm purchase is made with non-farm funds, enter it as a "check" drawn on the Non-Farm Funds account.**

   QuickBooks allows this because the Bank account type was used for setting up the Non-Farm Funds account.

   Choose Activities|Write Checks to open the Write Checks window. Be sure to select Non-Farm Funds in the Bank Account field at the top of the window—you don't want to make this entry in the farm checking account. Then record the expense details:

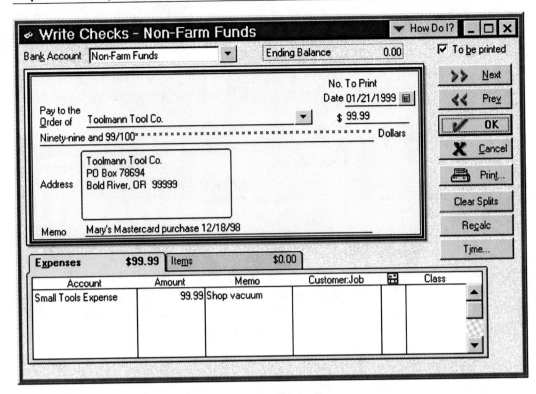

Understand that this entry *does not* correspond to a real check. That is, no check was actually written to Toolmann Tools; in fact, the Memo field tells us the purchase was made with a personal credit card. The Checks form just provides an easy way to enter purchases made with non-farm funds.

What about recording the addition to farm business capital which resulted from this purchase? Periodically you should make an adjusting entry to transfer the Non-Farm Funds balance to an equity account...we'll leave that discussion until later, in the Using a Non-Farm Funds Account topic in chapter 9.

Entering Purchases as Capital (Owner's Equity) Additions

Pros:

◆ Continuously maintains correct farm asset and equity account balances, without adjusting entries.

◆ Can be used by sole proprietorships and partnerships.

Cons:

♦ Requires General Journal entries and an understanding of debits and credits, so it's not easy for everyone.

♦ Should not be used by corporations.

This method accomplishes the same result as using a non-farm funds account but goes about it differently. Instead of using a special account to accumulate farm purchases made with non-farm funds, each purchase is entered directly as a capital addition to the farm business. This maintains correct asset and equity account balances all the time, so no adjusting entry is ever necessary.

Here are the basic steps involved:

1. **Open the General Journal.**

   Choose Activities|Make Journal Entry.

2. **Enter the farm purchase(s) made with non-farm funds by** *debiting* **the appropriate farm expense accounts and** *crediting* **an equity account to reflect the addition of capital to the farm business.**

   As the following example shows, you can enter any number of expense lines, and offset all of them with a single line posted to an equity account:

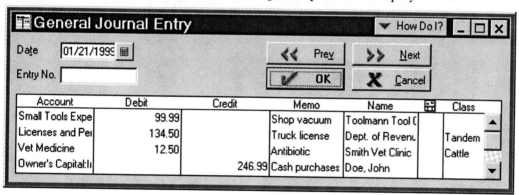

Debiting the expense accounts increases their balances. Crediting the Owner's Capital:Investments account increases its balance also. So this entry records both the farm expenses and the increase in farm business equity.

How should a partnership use this method? Partnerships should use separate equity accounts for each partner, which will allow properly keeping track of how much capital each partner has contributed to the farm business by making farm purchases with his or her own funds.

Corporations should *not* **use this method.** The only way stockholders can add to their ownership interest (equity) in a corporation is to purchase stock.

When a stockholder uses personal funds to buy something for the corporation, he or she should be reimbursed directly from farm business funds, such as with a check drawn on the farm bank account.

Reimbursing the Purchaser from Farm Business Funds

Pros:

♦ Keeps personal and farm business finances separate, and documents that separation by providing a good "paper trail" of transactions.

♦ Can be used by corporations.

Cons:

♦ Generates more paper work than the other methods—entering, printing, and depositing reimbursement checks, etc.

Only one step is involved:

1. **Issue a check or other payment to reimburse those who make farm purchases with their own funds.**

   Here's a check to Mary Doe, drawn on the farm checking account, for a farm purchases she made.

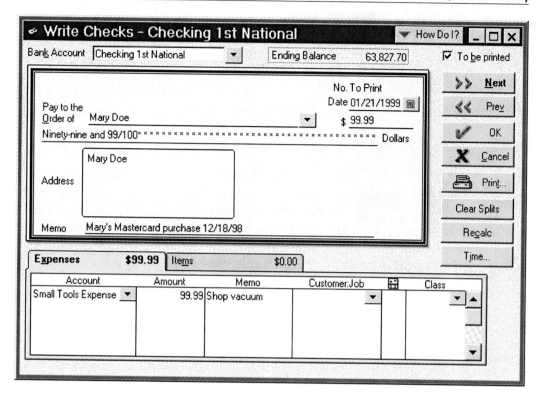

Reimbursements are simple and obviously keep farm and personal finances separate. They also work for any form of business organization. For corporations they're the only choice that maintains an adequate separation of corporate and personal finances.

Leave a good paper trail...

Especially if your farm business is a partnership or corporation, it's important to always leave a good "paper trail" describing a reimbursement. It's best if both the farm business and the person who made the purchase keep a detailed bill listing the purchased items. Be sure you have enough detail to show an IRS auditor or a court of law that the check you were issued from the farm business was a reimbursement and not a withdrawal of corporate or partnership funds.

Working with Credit Cards

Problem

How do I keep track of farm purchases made by credit card?

Solution

Use QuickBooks' special features for managing credit cards.

Discussion

QuickBooks helps you enter, organize, and reconcile credit card charges and payments in much the same way as working with a checking account, which makes the job easy.

Less easy, is deciding whether to include *personal* credit cards in the farm chart of accounts as if they were *farm* liabilities. Why would you want to? For easy entry of farm expenses paid by credit card, and to use QuickBooks' credit card management features without the need to set up a separate QuickBooks company for personal transactions. A personal credit card can be set up as a farm liability account if you're willing to (1) consider it as belonging to the farm business (even if it's been issued to an individual—to "John Doe" instead of "John Doe Farms"), and (2) use it strictly for farm purchases. For a discussion of the pros and cons see the shaded box titled "Personal Credit Cards as Farm Liabilities?", below.

It is possible to keep *separate* farm and personal expense records in *one* QuickBooks company file. See "Keeping Non-Farm Income & Expense Records in the Farm Accounts", on page 251.

The other approach is to set up a separate QuickBooks company file for personal spending, and enter credit card charges there. This approach only makes sense if you have a personal checking account that's separate from the farm account, and if you write checks from it to pay the credit card charges. From a farm records perspective, farm purchases made with a personal credit card handled this way are just like farm purchases you make with personal cash or other non-farm funds. See "How to Enter a Farm Purchase Made with Personal (Non-Farm) Funds" on page 245.

Because working with credit cards is very similar to working with checks or bills, the remainder of this section provides only brief examples and discussion. And, it assumes you are entering transactions for credit cards *maintained by the farm business*—regardless of whether they were actually issued to an individual or to the farm business.

Personal Credit Cards as Farm Liabilities?

Standard accounting practice requires that a set of financial records represent a single business entity. Technically speaking then, a QuickBooks company file for a farm business should only include *farm* assets and *farm* liabilities. Personal assets and liabilities—things like a house, car, personal investments, and personal debt—should be excluded.

So it might seem that personal credit cards (issued to an individual rather than to the farm business) should always be excluded from the farm business accounts. But not so. Here are two situations in which "bending the rules" a bit by including personal credit cards as farm accounts makes sense:

1. When a personal credit card is made available for *exclusive* use by the farm business.

   In this case the personal credit card is essentially loaned to the farm business, and is no longer used for personal purchases. All purchases made with the card are farm purchases and all payments are from farm funds, so there's no distortion of farm liabilities (there would be if the card were used for personal purchases). In reality, the card is used as if it had been issued to the farm business rather than to an individual.

   This approach works both for sole proprietorships and for partnerships. Purely from an accounting standpoint, it should also work if the farm business is a corporation. However, the corporation may not shield individual owners from personal liability if personal and corporate finances appear to have been mingled, so consult an attorney before you try it.

2. When a personal credit card is made available for use by the farm business, but also continues to be used for personal spending.

   As in the situation described above, the credit card is essentially loaned to the farm business and is used as if it were issued directly to the farm business rather than to an individual. Also as above, all credit card payments are made from farm business funds—usually the farm checking account. But here's the difference: the card is also used for *personal* purchases, and they are entered as owner withdrawals of capital from the farm business (by assigning equity accounts when those purchases are entered).

   If used carefully, this technique won't materially distort farm liabilities. And it works in most sole proprietorships and a few partnerships, but should never be attempted in a corporation because of the need to keep personal and corporate finances separate.

When do personal credit card purchases distort farm liabilities? If you normally carry an unpaid balance on the personal credit card, farm liabilities will be overstated by the credit card balance you are carrying at the time. But if you completely pay the credit card balance each month, there should be no distortion in the farm balance sheet except at those times when you have entered personal credit card charges in QuickBooks but not yet paid them.

I'm a cash basis record keeper...should I keep track of credit cards in QuickBooks?

Yes! Just be sure to select the Cash Basis report customization option when you prepare reports. It causes accrual entries to be ignored for preparation of the report. (Farm credit card purchases which have been entered in QuickBooks but not yet paid are accrual expense entries and should be excluded from cash-basis reports.)

How often should I enter credit card charges? ...weekly? ...once a month?

Regardless of how often you enter credit card transactions, try to develop a regular routine or cycle for entering them. That will help prevent errors such as duplicate entries of the same charge.

Most people wait until they receive their monthly credit card statement, then enter charges and credits directly from the statement. This makes it easy to compare their charge tickets against the amounts shown on the statement. It can also make reconciling the credit card account easier: the cleared charges and credits (the ones listed on the statement) will be the only ones listed in the Reconcile Credit Card window, which makes it easier to get the right ones selected!

But there's also nothing wrong with entering credit card charges from time to time during the month. This can help you avoid overspending the card's credit limit by staying aware of its outstanding balance, and keeps farm liability balances more current. Also, fewer credit card transactions will need to be entered at month's end before reconciling the account.

Here's a typical cycle for managing a credit card account in QuickBooks:

Make purchases with credit card
throughout the month

Receive monthly credit card statement

Enter charges and credits on the Credit Card
Charges form, matching statement charges with
charge tickets. Enter **farm purchases** as usual.
Enter **personal & non-farm purchases** as withdrawals of farm capital (owner's equity)

Reconcile the credit card account and
enter a bill for payment before the due date

Prepare payment using the Pay Bills window

Send payment to the credit card company

Setting Up a Credit Card Account

Set up a separate QuickBooks account for each credit card, just as you would for
different bank accounts.

To Set Up a Credit Card Account:

1. Open the Chart of Accounts window.

   Choosing Lists | Chart of Accounts is one way to do this.

2. Open the New Account window.

   Either type Ctrl-N, or click on the Account button in the lower part of the
   Chart of Accounts window and then select New from the pop-up menu.

   The New Account window opens.

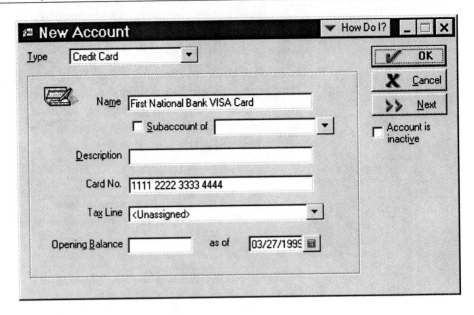

3. **Fill in the fields as desired.**

Here are comments on some of the important fields:

Type
Select the Credit Card account type.

QuickBooks handles accounts of the Credit Card type as current liability accounts.

Name
Use a more imaginative name than "Credit card". The name should uniquely identify the specific card you are working with. Because you may have more than one card from the same card company (Visa, MasterCard, etc.) it's a good idea to include the sponsoring bank's name in the account name, such as "First National Bank VISA".

Card No.
(Optional.) You may enter the credit card number here for reference purposes.

Tax line
Do *not* choose a tax line—they mostly apply to income and expense accounts.

Opening Balance
If the credit card currently has an unpaid balance, enter it here. It will appear as the beginning balance the first time you reconcile the credit card account in QuickBooks. For a new credit card the beginning balance should be zero.

4. **Click OK to save the new account.**

How to Enter Credit Card Charges

1. **Choose Activities|Enter Credit Card Charges.**

   QuickBooks opens the Enter Credit Card Charges window.

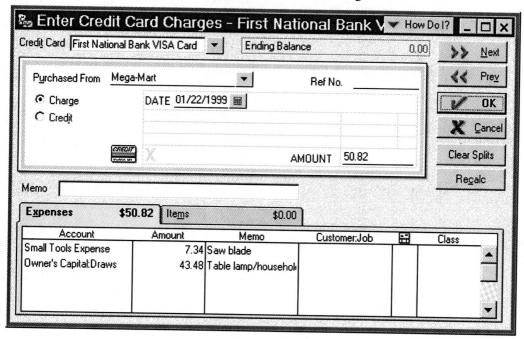

This window lets you enter either charges or credits. Note the "Charge" and "Credit" selections in the upper left part of the window. If "Credit" is selected, select on the "Charge" option to enter charges.

2. **Fill in the fields as desired.**

   Here are notes about important fields on the Credit Card Charges form:

   Credit Card
   If you have more than one card, be sure to select the correct one here.

More than one credit card account? Color-code them!

Color-coding accounts in QuickBooks helps reduce errors. Here's how:

1. Select a credit card account in the Chart of Accounts window.

2. Open the Enter Credit Card Charges window or a Register window for the account.

3. Choose Edit|Change Account Color and select the color you want to associate with the account.

The background color of forms and registers will be set to this color whenever you are working with this account.

Color coding works for any balance sheet account—bank accounts, loan accounts, etc.—providing visual confirmation that you're working with the account you intended.

Purchased From
Select the vendor's name here.

Expenses tab fields & Items tab fields
The lower area of the Credit Card Charges form works just like the lower part of the Checks form and the Bills form. For detailed information about individual fields, see the Entering Checks discussion on page 100.

The illustration shows two detail lines on the form's Expenses tab. One is for a saw blade—a farm expense—and the other is for a household (personal) purchase. Note that an equity account, Owner's Capital:Draws, is assigned to the household purchase so that it will be posted as a withdrawal of owner's equity from the farm business.

Accounting 101: Equity accounts to record personal spending

On QuickBooks forms that are normally used for entering expenses, such as the Credit Card Charges form or the Checks form, assigning an equity account to a transaction line records that line as a withdrawal of funds from the farm business. (That's what is happening when personal items are paid for with farm business funds.)

You can use a single equity account for all capital withdrawals and additions, but setting up several equity accounts lets you keep track of different categories of personal spending and income within the farm accounts. Here's an example of equity accounts set up for this purpose:

> Capital Additions
> > Wages
> > Other
>
> Capital Draw
> > Car expenses
> > Food
> > Household
> > Miscellaneous
> > Utilities

For more discussion see "Keeping Non-Farm Income & Expense Records in the Farm Accounts", on page 251.

Reconciling and Making Payment On a Credit Card Account

Reconciling credit card account(s) is an important part of the monthly cycle of accounting activities. Reconciling verifies the accuracy of your records. If you don't reconcile, you won't be aware of errors in your entries or of credit card charges you have paid due to mistakes by the credit card company or to credit card fraud.

When to Reconcile

It's best to reconcile a credit card account as soon as possible after receiving the monthly credit card statement and getting all of the month's charges and credits entered. If you find any errors or problems, that will give you more time to get them solved. (Most credit card companies require a dispute of any charge to be submitted in writing within 60 days of the statement on which the disputed charge appeared.)

How to Reconcile

The steps for reconciling a credit card account are mostly the same as for reconciling a bank account:

1. **Choose Activities|Reconcile.**

   The Reconcile window will open.

2. **In the Account to Reconcile field, select the credit card account you want to reconcile.**

   The Reconcile window's fields are slightly different depending on whether you are reconciling a bank account or a credit card account.

3. **Compare the amount in the Opening Balance field with the opening balance shown on your credit card statement.**

   QuickBooks uses the Ending Balance from the account's most recent prior reconcile as the Opening Balance for the current reconcile. So the Opening Balance field should match the opening balance on your credit card statement. However, if you've just created a new company file or if this is the first time you've reconciled a new credit card account, there may be reasons why the two don't match. Consult QuickBooks Help for ideas on what to do.

4. Enter the ending balance shown on your credit card statement in the Ending Balance field.

5. Enter Finance Charges, if any, in the "Transactions to be added" area.

   Don't forget to assign an expense account where the finance charge is to be posted.

6. Mark all of the "cleared" transactions—the ones that appear on the credit card statement.

   Scroll through the lists of Payments and Credits, and Charges and Cash Advancements, clicking on transactions that match those listed on the credit card statement. A ✔ appears to the left of transactions you've marked as cleared.

   If some transaction amounts don't match, correct the transactions. If you find transactions on the credit card statement that have not yet been entered in QuickBooks, open the Enter Credit Card Charges window and enter them. Then return to the Reconcile window. It is automatically updated to include the new transactions, which you may then mark as cleared.

An easy way to view or edit a transaction listed in the Reconcile window is to double-click it, or highlight it and then click on the Go To button.

7. When you have finished marking cleared transactions, look at the Difference displayed near the bottom of the Reconcile window.

   If the Difference is *not* 0.00 the account does not balance for the period of time covered by the statement. You need to either (1) find and fix the problem such as a missing transaction, an incorrect transaction amount, or an error on the part of the credit card company, or (2) adjust the account balance by the amount of the difference. Use this second option if you don't feel the amount of the difference justifies spending time to find it.

   To have QuickBooks automatically adjust the account balance, click on the Done button. If the Difference is other than zero, QuickBooks will ask if you want to allow the automatic balance adjustment. Click OK if you want QuickBooks to make the adjustment.

   If the amount is 0.00 you've successfully reconciled the account with the credit card statement. When you click Done, QuickBooks will ask if you want to write a check to pay the credit card now, or enter a bill to pay it later. In either case, QuickBooks fills in the basics of the check or bill, including the credit card account. You only need to enter a dollar amount (if you wish to pay other than the full amount due), and click OK to save the check or bill.

By the way, entering a check this way (as you finish reconciling) is no different from directly entering a check on your own. It's just that QuickBooks helps automate the job by mostly filling in the check.

Paying your credit card: enter a check or a bill?

If the due date is near, write a check—it takes care of making payment in a single step. But if you want to wait a few days to pay, entering a bill is best. That "calendarizes" the payment: the bill gets listed in the Reminders window, to help you get it paid on time.

Paying on a Credit Card by Check

Most of the time you probably make a credit card payment after receiving a monthly credit card statement, but you can pay on a card's balance at any time. For instance, maybe you need to pay down the balance to allow using the credit card for a large purchase such as a new computer.

Entering a check on your own to pay on a credit card is basically the same as entering a check immediately following reconciling the credit card account (described in the preceding section). The difference is that you have to fill in the check form yourself.

1. **Choose Activities|Write Checks to open the Write Checks window.**

2. **Fill in the Check form.**

   Complete the fields as you would for any check, but *be sure to select the credit card account* in the Account field.

3. **Print the check.**

   You may print the check by clicking on the Print button, or select the "To be printed" field and print the check later with a batch of other checks.

4. **Click OK to save the check entry and close the Write Checks window.**

5. **Mail the check.**

   When sending a check without a payment stub (the tear-off part of the credit card statement that you normally return with a payment) it's a good idea to print the credit card account number on either the check or a piece of paper accompanying the check.

Paying on a Credit Card with Another Credit Card

Unlike paying by check, paying on a credit card with another credit card involves no paperwork in QuickBooks—no check to print, etc. All you need to do is record the transfer of funds between the two accounts, which is easily done with the Transfer Funds form.

1. **Select Activities|Transfer money.**

   The Transfer Funds Between Accounts window opens.

2. **Fill in the fields, as appropriate.**

   Select the credit card you're making payment from as the Transfer Funds From account, and the card receiving the payment as the Transfer Funds To account. Also enter the date and the amount of the transfer. Here's an example:

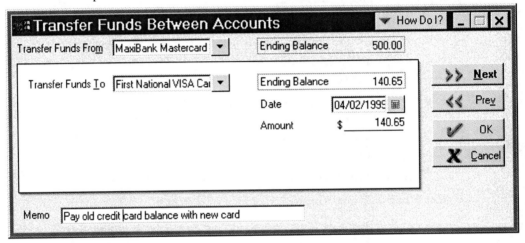

3. **Click OK.**

This transfer will appear in the Register window and the Reconcile window of both credit card accounts.

CHAPTER 6

Farm Income & Receivables

This chapter introduces QuickBooks forms and techniques for entering farm income and amounts others owe to the farm business (accounts receivable).

Where to Enter Farm Sales/Income ...in a Deposit? ...Cash Sale? ...Invoice?

Problem

QuickBooks has three different forms where I could enter income: a deposit, a cash sales form, or an invoice. When should I use each of these, and why?

Solution

Your choice depends on several factors, described in the sections below.

Discussion

In QuickBooks, you need to separate the idea of recording *income* from the idea of recording the *deposit* of that income to your bank account. The QuickBooks form you use determines whether these two steps are accomplished together (with one entry) or separately.

The form also determines whether you can enter quantity information with the transaction (the number of bushels or head or pounds sold), and whether or not you will use customer accounts to keep track of amounts you are owed (accounts receivable).

When to Use a Deposit

Use the Make Deposits window to record income when you don't need to enter a sales quantity.

Creating a deposit is the simplest way to record income. It's also very familiar to anyone who has moved to QuickBooks from a checkbook-based system like Quicken or from a manual ledger book. But the Make Deposits window doesn't give you any place to record quantities such as the number of bushels, pounds, or head sold. If you want to record a sales quantity, use either the Cash Sales form or the Invoice form.

The ManagePLUS add-on, described in chapter 7, allows **entering quantities in deposits**, which lets you avoid using Cash Sales and Invoice forms if desired.

When to Use the Cash Sales Form

Use the Cash Sales form when you want to record the quantity of things you have sold as well as the income, but you don't want to maintain customer accounts.

QuickBooks' Cash Sales form has a Qty (quantity) field where you can enter the quantity of items you've sold, such as a number of bushels, pounds, tons, or head. This is the main advantage of entering a cash sale instead of a deposit, since the Make Deposits window doesn't give you a place to enter quantities.

The Cash Sales form is intended for recording income you've actually received—when you have a check or cash in hand rather than being owed by someone. Transactions entered on the Cash Sales don't go through QuickBooks's customer accounts, as they would if you record income using an invoice.

When to Use an Invoice

If you make many sales on credit or want to be able to send statements of account to your customers, use invoices to record those sales.

Recording sales on the Invoice form is an absolute must if you want Quick-Books to track the amounts you are owed by purchasers of your products, or if you want QuickBooks to print statements to send to your customers.

But if you make very few credit sales—true of most farm businesses—the choice of whether to use invoices is less clear. Recording sales by using invoices in QuickBooks *requires* that you manage customer accounts in QuickBooks. Entering an invoice creates an Accounts Receivable balance for the customer. So when the customer pays, you must match the payment with the customer's account to offset the receivable balance.

Though QuickBooks makes working with customer accounts as easy as possible, for most users it requires additional learning and some trial-and-error effort. Depending on your understanding of customer accounts, it may be easier to avoid them (by not using invoices) and keep track of amounts you are owed by some other means, such as pending cash sales (page 160).

Here are the pros and cons of using invoices:

Reasons to use invoices:

♦ If you make a lot of credit sales, entering them as invoices is the easiest way to keep track of what you are owed by individual customers.

♦ If you want QuickBooks to print customer statements you must use invoices. Transactions entered on the Cash Sales form *are not* included when Quick-Books prepares customer statements.

♦ If you keep accrual accounting records you should use invoices to record sales on credit. By definition, accrual accounting requires that you record income when it is *earned* rather than when payment is received, and that's what happens when you use an invoice to record a sale on credit.

♦ Even if you keep cash-basis accounting records using invoices can still be worthwhile. With QuickBooks keeping track of amounts you are owed, a total for accounts receivable will be available when you want to prepare a balance sheet.

Reasons not to use invoices:

♦ You must understand how to manage customer accounts in QuickBooks, including how to use invoices and credit memos. Though not difficult to learn, it's unfamiliar territory for many farmers using QuickBooks.

♦ Using invoices requires more steps than merely depositing income when you receive payment.

♦ Problems can arise if you don't use invoices consistently. A common mistake of people who seldom use invoices is to simply deposit the income when a customer makes payment on an invoice, rather than using the Receive Payments window (Activities|Receive Payments). The typical result is that income is recorded twice and the customer's account doesn't get credited for the payment.

♦ Entering pending cash sales is a simpler alternative for many QuickBooks users. It gives cash-basis record keepers a simple way to use QuickBooks for keeping track of amounts they are owed, without the complexities of managing customer accounts. (See page 160 for details.)

Entering Sales the Simple Way, as a Bank Deposit

Problem

How do I enter a bank deposit?

Solution

Use the Make Deposits window.

Discussion

Using the Make Deposits window is the simplest way to enter deposits to your bank account. However, it doesn't give you a place to enter quantities such as bushels, weight, or number of head sold. For that you need to use either the Cash Sales form or Invoice form (both are described later in this chapter).

Before entering a deposit you must set up a QuickBooks account to represent your bank account, as discussed in chapter 3.

Though the example in this section deals with recording a sale, you will use the Make Deposits form for recording *any* deposit to your bank account including transfers from other accounts, sales of farm assets (machinery, livestock, land), and so on.

The ManagePLUS add-on, described in chapter 7, allows **entering quantities directly in the Make Deposits window.**

When to Enter a Deposit

Depending on which accounting system you used before, you may be in the habit of entering deposits "after the fact"—after you've already sent them to the bank. With QuickBooks, a better habit is to enter deposits *when you are preparing them to send to the bank.* Besides helping to keep your records current, this habit may save you some effort: you can let QuickBooks print a detailed listing of checks to accompany the deposit slip, rather than handwriting them all on the deposit slip.

How to Enter a Deposit

1. **Choose Activities|Make Deposits.**

    The Make Deposits window will open. *If you've entered cash sales or received payment on an invoice another window may appear first—see "What else might happen...", farther below.*

Here's a completed deposit example which records the deposit of a check received for selling soybeans:

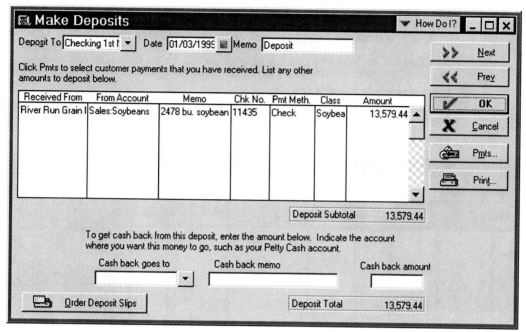

2. **Fill in the fields as desired.**

Here are notes about some important fields:

Deposit To
Be sure to choose the correct bank account if you use more than one.

Received From
This is the person or business who is the source of the deposited funds. Though QuickBooks may allow you to leave this field blank, don't! Get into the habit of entering names whenever requested in QuickBooks, to preserve the detail of your transactions.

From Account
Select the account that represents the source of the deposit, often an income account such as Grain Sales.

Class
(Optional.) Select a class to identify a deposit line with a particular enterprise such as Soybeans:1998.

Cash back goes to (QuickBooks 99 and later)
If cash is withheld from the deposit, this field lets you assign an account to

the withheld cash—usually a petty cash account or owner's equity/drawing account.

Cash back amount (QuickBooks 99 and later)
If cash is withheld from a deposit, enter the amount of cash withheld here.

3. Click OK to save the deposit and close the Make Deposits window.

What else might happen...

Sometimes when you open the Make Deposits window, the Payments to Deposit window (Undeposited Funds window in older versions) may open first:

Payments to Deposit					▼ How Do I?		✗
Select the payments you want to deposit, and then click OK.						✓	OK
✓	Date	Type	No.	Pmt Meth	Name	Amount	
	01/03/1999	RCPT	1		River Run Grain F	5,567.60	✗ Cancel
							? Help
							⋏⋏⋏ Select All
							⋏⋏⋏ Select None

This happens if you've previously entered cash sales or received payment on an invoice and selected the "Group with other undeposited funds" option. Payments listed in the Payments to Deposit window are from the Undeposited Funds account—a sort of "holding tank" for checks and cash payments you have received and recorded as income but not yet deposited. This window opens automatically to let you select payments to include in the deposit.

Here's what to do:

1. Select any undeposited payments you want to include in the deposit, by clicking on them.

A check mark will appear to the left of those you have selected.

2. Click OK.

The Payments to Deposit window will close, and the payments you selected will automatically be included in the deposit.

Printing Deposit Details

Most banks require a detailed deposit slip listing the checks and other funds included in a deposit. Once you have entered a deposit, QuickBooks can print this listing for you: just click on the Print button in the Make Deposits window.

Your bank may still require that you include their standard magnetic-ink deposit slip. But some banks will also accept your QuickBooks-printed check listing as a source of deposit detail.

 QuickBooks 99 and later versions can **print deposit details directly to magnetic-ink deposit slip forms** acceptable to most banks, letting you avoid handwriting deposit slips entirely.

Entering Deductions (Yardage, Hauling, Commissions, etc.) from Bank Deposits

Problem

Often the checks I receive are for a net amount—the amount left over after deducting commissions, yardage, hauling, etc. I'd like to record these expenses separately from the income, so I can keep information on the cost of each type of deduction. How can I enter deductions separately from income in a deposit?

Solution

Enter the gross amount of income on one line of the deposit and the deductions on other lines, as negative amounts.

Discussion

QuickBooks allows lines with negative dollar amounts in deposits, so long as the deposit total is positive. (The same is true in cash sales receipts and invoices.) In a deposit, *negative* dollar amounts assigned to expense accounts will appropriately *increase* the balance of those accounts.

How to Enter Deductions (Expenses) in a Deposit

Here's an example deposit of a livestock check, with negative lines entered for various deductions:

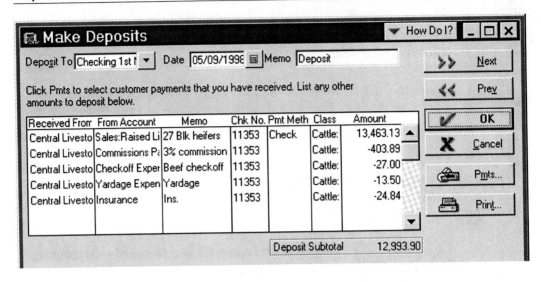

Withholding Cash from a Bank Deposit

Problem

When part of a deposit is withheld as cash, how do I show that when I enter the deposit?

Solution

Enter a negative line in the deposit for the amount of cash withheld. Or, if you have QuickBooks 99 or later, enter the cash withheld in the special "Cash back" fields provided near the bottom of the Make Deposits window.

Discussion

In both examples below, a grain sales check is deposited and $100 of the deposit is withheld as cash. The results of both transactions are: (1) the full income amount is assigned to the grain sales account, (2) the cash withheld is recorded as a withdrawal of owner's equity from the farm business, and (3) the net amount of the deposit gets added to the checking account balance.

You may select any owner's equity account for cash withdrawals. The Owner's Capital:Draws account used in these examples illustrates the idea of keeping track of capital withdrawals separately from capital additions to the farm business.

Negative Deposit Line Example (QuickBooks 6.0 & Earlier)

This is the Make Deposits window from QuickBooks 6.0. Other versions are arranged differently, but the entry would be basically the same:

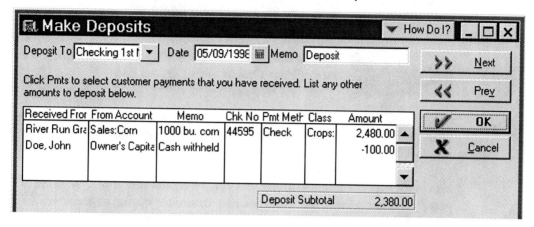

Cash Back Field Example (QuickBooks 99 & Later)

In later QuickBooks versions, cash withheld from a deposit can be included in the "Cash Back" fields at the bottom of the form, where it's not necessary to use a negative dollar amount.

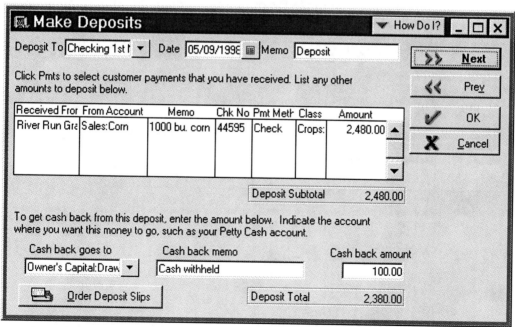

Income from Refunds & Rebates

Problem

I canceled a truck insurance policy and the company issued a refund check. How do I properly enter the refund?

Solution

Include the refund check in a deposit and assign the insurance expense account to that deposit line.

Discussion

Entering refunds for returned products or unused services is easy. When you deposit them, just assign the same account as you would for recording them as an expense. This will cause QuickBooks to subtract the refund from the balance in the expense account. Think of it as entering the refund as a "negative expense". Product rebates, like those you may receive for buying a specific herbicide or livestock parasite pour-on, are entered the same way.

Would you rather show refunds and rebates as income? If so, set up an income account, name it something like Cash Discounts & Rebates, and use it instead of assigning an expense account.

Here's an example deposit of two checks. One is a refund for a canceled insurance policy on a farm truck. The other is a rebate received for buying a livestock parasite pour-on.

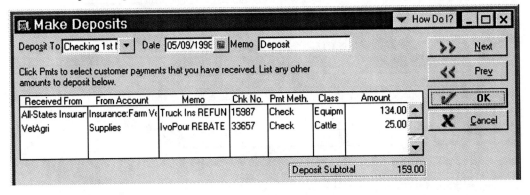

Recording Sales Dollars and Quantities: The Cash Sales Form

Problem
How do I enter quantity information (bushels, pounds, or gallons sold) along with income, for the things I've sold?

Solution
Use a Cash Sales form when you want to record a sales quantity along with the income you have received.

Discussion
Don't let the "cash" in Cash Sales confuse you. You may use this form to record any sales income that will be deposited in the farm checking account. Most often this income will be in the form of checks, but it may also be currency (cash).

How to Enter Income on the Cash Sales Form

1. **Choose Activities | Enter Cash Sales**

   The Enter Cash Sales window will open. Here's an example, with some corn sale transactions entered:

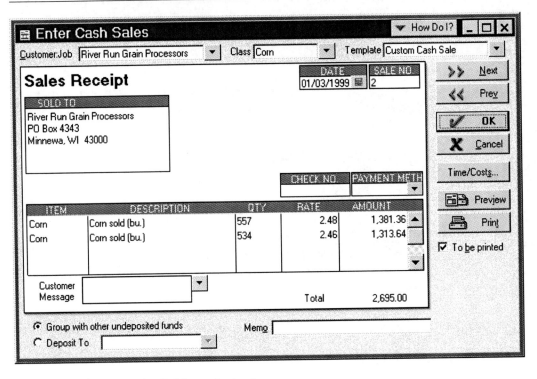

2. **Fill in the fields as desired.**

Here are notes about important fields and features of the Cash Sales form:

Class
The importance of using Classes for enterprise information was discussed in chapter 3. If you want to use Classes but don't see a Class field, you need to turn on Class tracking in the Preferences window (see page 25).

Template
This field gives you access to commands for customizing the Cash Sales form. Click the down arrow in the Template field, then choose the Customize option. After that, you'll be able to choose the fields you want shown on the form, change the layout of printed versions of the form, and so on.

Item
Here you must select a QuickBooks item you have created to represent the type of income you are entering.

If no appropriate item exists, type in a new item name. When you try to move to the next field, QuickBooks will tell you the name is not in the items list and ask if you want to add it as a new item. See chapter 3 for information on setting up items.

Qty (quantity)

This is where you enter the sales quantity—the number of bushels, pounds, tons, or head sold. This field provides one of the main advantages for using the Cash Sales form instead of a deposit. (The Make Deposits window does not give you a place to enter sales quantities.)

Rate

This field shows the price per unit for each item, such as the per-bushel price of grain, and is an optional entry. Normally you don't need to type anything here. Just enter the Qty (quantity) and Amount, and QuickBooks will automatically calculate and fill the Rate field for you.

Amount

You will usually enter the dollar amount for the transaction line here. But if you prefer, sometimes you may enter the Qty and Rate and let QuickBooks calculate the Amount for you.

How the Qty, Cost, and Amount fields are related

In any Checks or Bills form you only need to enter information in two of these fields—or just one field if it's the Amount field. Here are the possible combinations:

Qty and Cost - Fill these fields and QuickBooks will calculate and fill in the Amount field.

Qty and Amount - Fill these fields, and QuickBooks will calculate and fill in the Cost field.

Amount - QuickBooks will accept a transaction with just the Amount field filled, if that's how you want to enter it.

These three fields are always linked in this way, so be careful when you edit transactions. For example, suppose you originally enter the Qty and Amount on a transaction line and QuickBooks calculates and fills in the Cost field. If you later go back to that transaction line and change the number in the Cost field, the Amount will be recalculated automatically...so be sure that's what you want to do.

Group with other undeposited funds / Deposit to

These options (in the lower right corner of the Enter Cash Sales window) give you control over how and when to deposit income from the cash sale. Choosing "Group with other undeposited funds" will cause the income to be posted to an account called Undeposited Funds, a sort of "holding tank" for income you've received but not yet deposited. Choosing "Deposit to", and selecting a bank account in the field immediately to the right, will cause the income to be posted ("deposited") directly to that bank account.

Most of the time the "Group with other undeposited funds" option is best. It lets you accumulate received payments until you are ready to include them in a deposit. The "Deposit to" option automatically enters a separate deposit for *each* cash sale, so it tends to clutter the checking account's register with deposit entries.

3. Click OK or Next to save the cash sale entry.

An alternative to the Cash Sales form...

Only the Cash Sales and Invoice forms allow entering sales quantities. So you must normally use one of those two QuickBooks forms to associate quantity information with sales. But the ManagePLUS add-on for QuickBooks, described in chapter 7, lets you enter **quantity information directly on most QuickBooks forms,** including deposits. So it lets you bypass using the Cash Sales and Invoice forms if you want.

ManagePLUS also supports entering **two quantities** in transactions—something neither the Cash Sales no Invoice forms allow.

Entering Deductions (Yardage, Hauling, Commissions, etc.) on Cash Sales Forms & Invoices

Problem

Often the checks I receive are for a net amount—the amount left after deducting commissions, yardage, hauling, etc. But I want to separately record the income and the deductions, so I can know how my costs for hauling and other such expenses. How can I do this in the cash sales form?

Solution

Enter the gross income on one line of the Cash Sales (or Invoice) form and the deductions on other lines. Assign *negative* dollar amounts to the deductions, so the net amount of the cash sale will equal the amount of the check.

Discussion

QuickBooks allows negative lines on Cash Sales and Invoice forms, so long as the form's net total is positive.

The example below is for a cash sale but would work identically for an invoice.

Cash Sale Deductions Example

Entering deductions from income on cash sales receipts and invoices is an easy two-step process:

1. **Set up a QuickBooks item for each type of deduction you may want to enter, assigning it the Discount item type.**

The Discount item type is for entering negative amounts (deductions) on QuickBooks sales forms. Here is how the New Item window might look for a Commissions deduction item.:

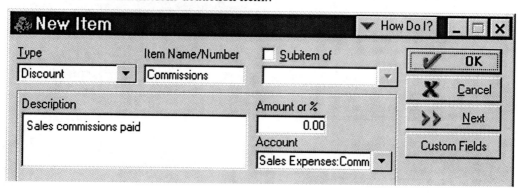

If you want profit and loss reports to show separate totals for each type of deduction, then assign a different expense account to each item you set up for deductions. Here are examples of expense accounts which would provide separate totals for several different kinds of sales deductions (assuming each were assigned to a different item):

> Sales Expenses
> Checkoff
> Commissions
> Hauling
> Insurance
> Yardage

(By the way, the Account field in the illustration is too short to show the full account name, which is Sales Expenses:Commissions.)

2. **Use the deduction items you have set up when adding deductions (negative lines) to Cash Sales or Invoice forms.**

This example shows how deduction items may be used on a Cash Sales form:

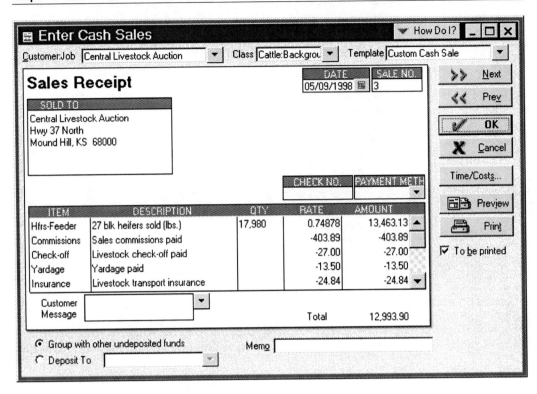

A Simple Way to Track What You're Owed: "Pending" Cash Sales

Problem

I don't want the complication of managing customer accounts, yet I'd like to use QuickBooks for keeping track of amounts others owe me. Is there a way to do this?

Solution

When you've sold something without receiving payment, create a cash sales entry and mark it as Pending to prevent it from posting as income. Later when you are paid, go back to the cash sales entry and mark it as Final to post the income to your accounts.

Discussion

Marking cash sales as Pending is a simple but effective way to keep track of amounts you are owed. This technique works well if you keep cash basis records because income is not posted until payment is actually received. You can even print a copy of the pending cash sale as a bill to send to the person who owes you payment.

Using **pending cash sales** to keep track of amounts owed to the farm business is proper accounting practice **only for cash basis record keepers.**

If you keep accrual records, you need to keep track of amounts owed to the farm business as Accounts Receivable (by entering those amounts on the Invoice form). Pending cash sales don't work for this because they are not posted to any accounts and therefore will not be included if you prepare an accrual-basis balance sheet.

As a practical matter though, many farm businesses prepare balance sheets outside of QuickBooks, using a spreadsheet or some other software package. If that's what you do, there's no reason to shy away from using the pending cash sales technique described here. Just remember to include the value of those pending cash sales as an asset when preparing a farm balance sheet.

How to Mark a Cash Sale as Pending

1. **Create the cash sale entry.**

   Enter a cash sale just as you would if you had received payment for the sold item(s), as described earlier in this chapter.

2. **With the Enter Cash Sales window still open, mark the cash sale as Pending.**

   Choose Edit|Mark Cash Sale as Pending. This keeps the cash sale entry "in limbo" by preventing QuickBooks from posting it to any accounts until later, when you mark it as Final (which removes the Pending designation).

3. **Click OK or Next to save the cash sale entry.**

How to Mark a Cash Sale as Final

You will normally mark a Pending cash sale as Final after receiving payment of the amount you were owed, to allow the cash sale to be posted as income.

1. **Open the desired Pending cash sale entry.**

   A good way to find a particular Pending cash sale is to first get a list of pending cash sales by opening the Pending Sales report window (Reports|Sales| Pending Sales), then double-click the desired cash sale to open it in the Enter Cash Sales window.

2. **Mark the cash sale as Final.**

Choose Edit|Mark Cash Sale as Final. You should also choose how funds from the sale are to be handled, by selecting either "Group with other undeposited funds", or "Deposit to".

3. **Click OK or Next to save the cash sale entry.**

Because it is now marked as Final, income from the cash sale will post normally.

How to Print Out a Cash Sales Form as a Bill/Invoice

You can use a printed copy of a cash sale entry as a bill to send to the person who owes you money. If you intend to do this, print the cash sale *before* marking it as Pending, because pending cash sales have "PENDING" printed in large letters diagonally across the form. Or, you may temporarily mark the sale as Final while printing it, as these steps describe:

1. **Find the cash sale entry you want to print.**

2. **Mark it as Final (you will change it back to Pending later).**

Choose Edit|Mark Cash Sale as Final.

3. **Change the title of the cash sales template to "Invoice", "Statement", "Bill", or whatever you want as a title on the printed copy.**

You may either change the title of the current template or create a new one. If you create a new template, be sure to select it in the Template field (in the upper right-hand corner of the Enter Cash Sales window) before printing.

Using Form Templates

Recent QuickBooks releases let you print the same form in different ways by creating multiple *templates* to control form appearance. If you frequently print out cash sales entries as bills, it's best to create a template specifically for that purpose. Here's how:

1. In the Template field in the upper right-hand corner of the Enter Cash Sales window, choose Customize.

QuickBooks will display a Customize Template dialog.

2. Click on the New button.

QuickBooks will open the Customize Cash Sale window, a set of tabbed dialogs containing the template's settings.

3. Enter a name for the template, and change other settings as desired.

Note that the template's name *will not* appear on the printed form. To change the report title—to "Invoice" or "Bill", for example—you must type a different title in the Default Title field.

4. Click OK to close the Customize Cash Sale window.

To apply the new template be sure it is the one selected in the Template field, in the upper right-hand corner of the Enter Cash Sales window. For more information about templates, consult the QuickBooks Help system.

4. Click on the Print button to print the cash sale form.

5. Re-mark the cash sale as Pending.

Choose Edit|Mark Cash Sale as Pending.

Re-marking the cash sale as Pending *reverses the posting of the cash sale income,* which occurred when the sale was marked as Final in step 2.

Need to Track Customer Accounts and Print Statements? Use Invoices

Problem

I run a custom harvesting business and want to keep track of which customers have paid and which haven't. I also want to send out a bill (invoice) for each job, and a monthly statement of account to those who still owe me. What's the best way?

Solution

If you enter the custom harvesting charges on the Invoices form, QuickBooks will keep track of what each customer owes you. You can print a copy of each invoice as a bill to give to your customers, and a monthly statement of account to send to those customers who still owe money.

Discussion

Invoices are for recording credit sales—sales for which you don't immediately receive payment. Invoices are also the only way to enter charges that will appear on statements of account printed from QuickBooks. Said differently, customer accounts are only maintained for charges entered as invoices.

An invoice is a document providing a detailed description of a sale of goods or services. It names the buyer and the seller, the date of the transaction, and the terms of sale (discounts, payment due date, etc.). Technically, it should be presented to the buyer when goods are transferred to the buyer's possession or services are performed. Just as often though, the buyer may not receive an invoice for several days, or maybe at month's end when a statement of account is received, or maybe not at all.

The monthly billing statement you get from your veterinarian or farm supply cooperative is often a good example of a "hybrid" document which is part invoice and part statement of account. Considerable detail about the items or services purchased is included, because, depending on practices of the seller, you may not necessarily have received copies of the individual invoices listed on the statement. (By the way, recent QuickBooks versions can also prepare this kind of detailed billing statement. See Statement Charges in the QuickBooks User's Guide.)

So should you always enter an invoice when you sell something and aren't paid immediately? If you keep accrual records the answer is "Yes", but if you keep cash-basis records the answer is "No" or at least "Usually not". Cash-basis accounting normally requires that income only be entered after payment is actually received. However, QuickBooks makes it possible to enter credit sales on invoices and still get cash-basis reports. Still, there are a couple of things you need to know before you consider using invoices:

1. **Using invoices in QuickBooks *necessarily* involves managing customer accounts.**

   Handling customer accounts requires more effort and understanding than simply entering income as a cash sale or deposit. If you only send out a few bills each year, working with customer accounts in QuickBooks may not be worth the learning required—especially if you are new to QuickBooks and already have plenty to learn.

2. **Using invoices in QuickBooks *necessarily* involves accrual accounting.**

   An invoice records income *as of the invoice date*, regardless of when payment is actually received. Cash-basis record keepers who use invoices must be careful to choose the Cash Basis report customization option to assure getting cash-basis reports. (This report option causes QuickBooks to ignore accrual transactions—such as invoiced sales for which payment has not been received—during report preparation.)

If you have just a few customers or rarely need to send out bills, you might try an alternative to invoices such as the pending cash sales technique described on page 160. But if any of the following apply to you, then you should be using invoices to record sales:

◆ You frequently need to send out customer statements.

◆ You have enough credit sales and/or customers that you need an organized way to keep track of the amounts you are owed.

◆ The time lag between making a sale and getting paid is typically long enough that Accounts Receivable can grow to a significant amount, and you want to stay aware of that amount and want it to be included on balance sheet reports.

Here is a flowchart of the typical steps involved if you use invoices:

Enter credit sales on the Invoice form

Give a copy of the invoice to your customer, or keep it to send later with the customer's statement

Assess finance charges and **print customer statements** (usually monthly)

Receive payments from customers and apply them to the appropriate invoices

Deposit the received payments

Watch for overdue invoices in the Reminders window, and follow up to collect payment

How to Work with Invoices, Customer Accounts, and Customer Statements

Here are some specifics about the flowchart of steps described above:

1. Enter an invoice when you sell something on credit—i.e., when you don't receive payment at the time of the sale.

   If you do receive payment at the time of the sale, you may use the Cash Sales form instead.

   To open the Invoice form, choose Activities | Create Invoices.

   Filling in an Invoice form is just like filling in the Cash Sales form. You select items to describe what was sold or what services were performed (like custom harvesting), and enter the quantity and price of each item. Items describe what you are selling and where to post the income (to which accounts). See page 68 for a discussion of items.

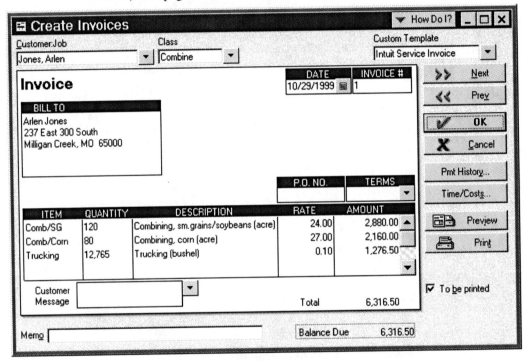

2. Print out a copy of the invoice to give to your customer, if you wish.

   Like many QuickBooks forms you may print invoices one at a time, by clicking the Print button on the Invoice form, or in a batch by (1) selecting the "To be printed" check box on individual invoices, then (2) printing the batch (File | Print Forms | Print Invoices).

3. Periodically you may assess finance charges on overdue account balances (if you choose to) and send out monthly statements.

   Most businesses send out customer statements once a month, but you may generate them at any time, or not at all if you prefer.

Choose Activities|Assess Finance Charges to apply finance charges on over-due customer accounts. Choose Activities|Create Statements to open the Select Statements to Print dialog, which lets you prepare statements for some or all of your customers:

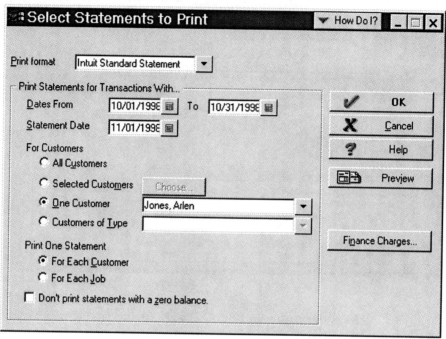

Consult the QuickBooks User's Guide or Help system for details on assessing finance charges and preparing customer statements.

4. **When a customer pays, apply the payment to the customer's invoices in the Receive Payments window.**

   This step is a critical part of managing customer accounts in QuickBooks. It matches the payment with the appropriate customer's invoices, and marks them as Paid. Depositing a customer's payment without using the Receive Payments window will not mark the customer's invoices as Paid—they will still be listed on the next statement you print for that customer.

   Open the Receive Payments window by choosing Activities|Receive Payments. Here's how the window looks when applying a customer payment to an outstanding invoice:

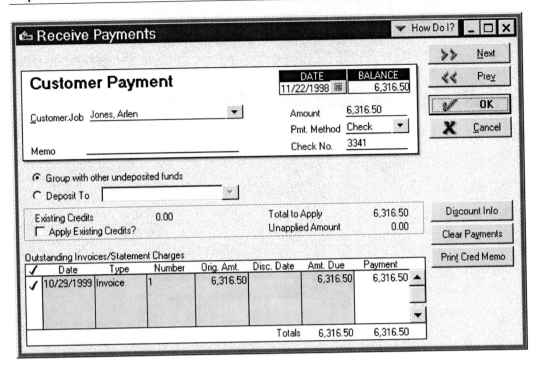

 The most common mistake when using invoices is to simply deposit a customer's payment without using the Receive Payments window. The result is that income from the sale gets recorded twice and the customer's account is not credited with the payment.

Consult the QuickBooks Help system for details on using the Receive Payments window.

5. **Deposit the customer's payment.**

   QuickBooks makes this step almost automatic. Whenever you open the Make Deposits window QuickBooks checks the Undeposited Funds account. If it finds any undeposited payments it displays them so that you may select any or all of them to include in the deposit entry.

6. **Watch for overdue invoices.**

   An important part of managing accounts receivable is to stay aware of those customers who are behind schedule in making payments. When an invoice is overdue, QuickBooks will display it in the Reminders window if you have set the proper Reminder preferences.

   Also, you can print the Accounts Receivable Detail report (Reports|A/R Reports|Aging Detail) to get a printed listing of current and past-due invoices.

These are just the basics of using invoices, but there's more to learn. Consult the QuickBooks manual or the Help system for details about managing customer accounts, including how to enter credit memos, assess finance charges, prepare billing statements, etc.

Entering Farm Income Received as Cash

Problem
Some of my hay customers pay with cash. How should I enter the cash income in QuickBooks?

Solution
The proper way to enter cash income depends on what you do with the cash. Do you deposit it in the farm checking account or do you keep it to spend for farm or personal items?

Discussion
As mentioned often throughout this book, keeping the farm's finances separate from personal finances is important, at least in accounting records. Farm income received as cash is one area where this can be difficult.

There should be no problem if you deposit the cash in farm checking—you include the cash income in a deposit entry just like any other farm income. But what if you keep the cash on hand to spend rather than depositing it? And what if you spend it on personal items instead of farm purchases? In that case your QuickBooks entries should reflect the withdrawal of owner's equity from the farm business.

Why keep farm and personal records separate?

Here are some of the reasons:

- Having accounting records specific to the farm business makes it easier to gauge farm profits.

- When calculating income taxes, expenses such as interest which are always deductible as farm expenses are not always deductible as personal expenses. (If your records blur the lines between farm and personal spending, some deductions may be disallowed by the IRS.)

- In partnerships and corporations it is important to treat the farm business as a stand-alone entity for proper distribution of profits to the owners/shareholders.

- Corporations which mingle business and personal (shareholder) finances risk losing some benefits of incorporation, such as limited liability for the shareholders.

However, there *is* a safe way to keep farm and non-farm income and expenses records in the same QuickBooks company file. See "Keeping Non-Farm Income & Expense Records in the Farm Accounts", on page 251.

The rest of this section shows various ways to enter farm income received as cash and keep farm and personal finances separate *even if you use the cash for personal spending!*

By the way, the discussion here is probably incomplete without a visit to the topic titled "The Best Way to Keep Track of Cash Spending & Cash Income", on page 231. It's a broad look at ways to streamline how you handle cash transactions, both for cash flowing into and out of the farm business.

The techniques discussed in this section generally apply any time cash is received by the farm business, whether the cash represents income, or the sale of a farm asset, or otherwise.

If You Deposit the Cash...

Cash payments you deposit in the farm checking account are entered in QuickBooks just like income received by check. You can enter them directly in a deposit (in the Make Deposits window), or enter the income first on the Cash Sales form and then include it in a deposit later. Either method assigns income to the appropriate account and increases the farm checking account's balance.

If You Add the Cash to the Farm's Petty Cash Fund...

Petty cash is a small amount of cash kept on hand for miscellaneous spending. If the farm business keeps its own fund of cash strictly for farm spending, set up a petty cash account in QuickBooks and enter the cash transactions there. See "Maintaining the Farm's Cash Fund (Petty Cash)" on page , for examples of setting up and using a petty cash account.

If You Use (Withdraw) the Cash for Personal Spending...

If the farm business receives cash but you keep it for personal spending, you are withdrawing funds (owner's equity) from the farm business. In this situation your QuickBooks entries need to accomplish two things: they must (1) record the farm income associated with the cash received, and (2) record the withdrawal of owner's equity. Below are two different methods for accomplishing both goals.

Method #1: Enter the cash income as a "deposit" to the Non-Farm Funds account...

For most people this is the easiest approach because no understanding of debits and credits is required. Farm income that you keep for personal spending gets recorded as a "deposit" to Non-Farm Funds, a special-purpose account set up for accumulating additions and withdrawals of farm capital (owner's equity). You'll find a full discussion of setting up and using a Non-Farm Funds account beginning on page 243, but here is a short example:

1. **Set up a Non-Farm Funds account in QuickBooks if you don't already have one.**

   Open the Chart of Accounts window, then choose Edit|New account. Choose "Bank" as the account type. Name the account Non-Farm Funds or some other name you prefer, and give it a beginning balance of zero.

2. **To record cash farm income, enter it as a "deposit" to Non-Farm Funds.**

   QuickBooks allows this because Non-Farm Funds was assigned the Bank account type.

   Choose Activities|Make Deposits to open the Make Deposits window. Be sure to select Non-Farm Funds in the Bank Account field at the top of the window—you don't want to enter the deposit in the farm checking account. Then enter the deposit:

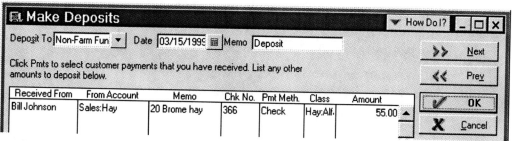

This "deposit" does not correspond to an actual bank deposit, of course. The Make Deposits form just provides an easy way to enter the cash income in the Non-Farm Funds account

As for recording the withdrawal of owner's equity from the farm business, that's done by periodically making an adjusting entry to transfer the Non-Farm Funds balance to an equity account. See "Using a Non-Farm Funds Account", on page 243, for details.

Method #2: Enter the cash income in the General Journal, and use an equity account as the offsetting account to record the owner's equity withdrawal...

This method accomplishes the same end result as using a Non-Farm Funds account but goes about it differently. Instead of accumulating personal withdrawals in a special account, each one is entered directly as an owner's equity withdrawal. This approach maintains correct asset and equity account balances all the time, so no transfer or adjusting entry is ever necessary. Here are the steps:

1. **Open the General Journal.**

   Choosing Activities|Make Journal Entry is one way to do this.

2. **Enter the cash income as shown in this hay sale example:**

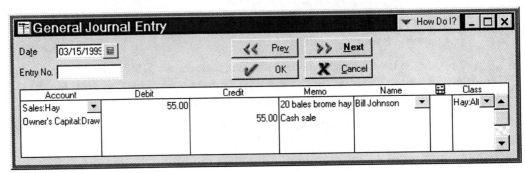

This entry credits the Sales:Hay income account and debits Owner's Capital: Draws to record the withdrawal of owner's equity. (Any equity account could have been used, but it makes sense to have one or more equity accounts set up specifically for tracking equity withdrawals separate from equity additions to the farm business.)

May a partnership use this method? Yes, but use a separate equity account for each partner, to properly record who withdrew the cash.

What about corporations? No. When a stockholder uses farm business funds for personal spending he or she should reimburse the corporation, to keep corporate and personal finances completely separate.

Another way to handle cash income is to simply pocket the cash without recording it. It's no secret that income received in small cash payments is grossly under-reported on Federal tax returns. But besides being illegal, omitting large amounts of cash income from the farm records can affect the profitability picture of the farm business or of a particular enterprise, as viewed by your lender, off-farm partners, or shareholders. It's your decision...

172

CHAPTER 7

Quantity Information in Income & Expense Transactions

One of the management functions of an accounting system is to collect and report quantities. Information like the number of bushels of crop sold, the pounds and number of livestock sold, the pounds of fertilizer applied, or the pounds of feed fed can be valuable for estimating production costs and projecting future input needs and expenses.

Including Quantity Information in Your Transactions

Problem

How can I include quantities in my QuickBooks transactions, like the number of bushels, pounds, or gallons of the things I buy and sell?

Solution

You can use the items feature to enter a quantity for each item bought or sold, but only on forms which allow using items (Checks, Bills, Cash Sales, Invoice). An alternative is the ManagePLUS add-on product, which permits entering quantity information on a wider array of QuickBooks forms and provides enhanced quantity reporting.

Discussion

One area where QuickBooks' developers have obviously scrimped, is on features for recording and reporting on quantity information. QuickBooks' quantity handling features were mostly built around the program's inventory system. But that inventory system is designed for handling retail business inventories and cannot be used for most farm business inventory types. As a result, the amount of quantity information available from QuickBooks is a disappointment to many farm business record keepers.

The biggest part of the problem is that QuickBooks has almost no capability to produce quantity *reports* for anything other than inventories. But the ManagePLUS add-on and recent feature additions to the QuickBooks Pro versions at least partly solve the problem.

For information about obtaining the ManagePLUS product, see Appendix A: Other Resources and Information.

Using Items to Enter Quantity Information

The only method QuickBooks provides for entering quantities is by using items, which are names you set up in the Items list to identify things you buy and sell. (See page 68 for information about setting up items.) Any QuickBooks form which supports using items also gives you a place to enter the quantity of each item purchased or sold.

Only a few QuickBooks forms support using items, the important ones being the Check, Bill, Invoice, and Cash Sales forms. The Check and Bill forms allow using items but don't require them (you can select accounts instead, if you prefer). Two tabs are visible in the lower part of the Check and Bill forms if the items feature is available:

Expenses	$0.00	Items	$0.00

If your Check and Bill forms don't have these tabs, you need to turn on Quick-Books' inventory features (page 28 tells how) to make items available.

Here are some expense entries on the Items tab of the Checks form. Note that a quantity is entered in the Qty field for each of the seed corn items. (For a discussion of the fields on the Items tab of the Checks form, see page 101.)

Expenses	$0.00	**Items**	$1,680.70				
Item	Description	Qty	Cost	Amount	Customer:Job		Class
Seed corn	Seed corn 6303	10	96.00	960.00			Corn
Seed corn	Seed corn 7330	10	75.50	755.00			Corn
Cash discnt	Cash discount		-34.30	-34.30			Corn

Select PO   Receive All   Show PO

The Invoice and Cash Sales forms don't have tabs, because they require using items for all entries. Here's an example of using items in the detail area of the Cash Sales form:

ITEM	DESCRIPTION	QTY	RATE	AMOUNT
Corn	Corn sold (bu.)	557	2.48	1,381.36
Corn	Corn sold (bu.)	534	2.46	1,313.64

For a discussion of fields on the Cash Sales form, see page 155.

Entering Quantity Information in Memo/Description Field

Several QuickBooks forms—Deposits, for example—do not support using items. On these forms it is impossible to enter quantity information unless you use an alternative to the standard QuickBooks approach, such as the Manage-PLUS add-on.

ManagePLUS allows entering quantity information in the Memo field (it's called the Description field on some forms) of any transaction. This means you can enter quantities on nearly all forms, including the Deposits form and the General Journal. ManagePLUS extracts quantity information stored in Memo fields to report quantity totals and other quantity-related statistics.

For ManagePLUS to extract quantity information, the information must be entered in a certain way. The rules are simple though: type the quantity at the beginning of the Memo field, and follow it with a blank space to separate it from other Memo text. Here is a deposit example which records income from corn and soybean sales and includes the number of bushels sold, in the Memo field:

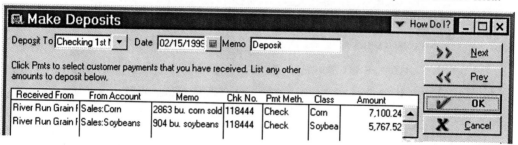

A side benefit of being able to enter quantities in the Memo field of transactions is that it can save a step or two in the data entry process. The deposit entry above is a good illustration of this. Including quantities directly in the deposit's Memo field lets you bypass the step of entering the income and quantities on a Cash Sales or Invoice form first, before entering the deposit. (The Cash Sale and Invoice forms allow entering quantities because those forms support using items.)

It's important to note that ManagePLUS *allows* storing quantities in the Memo field but does not require it. You may also enter quantities by using items if you wish, and can use either method at any time—you aren't limited to using just one or the other.

Especially if your business involves **livestock**, you may want to include **two quantities** in some transactions, such as both the weight *and* number of animals bought or sold. The ManagePLUS add-on lets you enter two quantities in most transactions and provides **totals for both quantities,** plus related statistics such as **average weight** per head and **average price**. See page 179 for more information.

Getting Quantity Reports

At the time this is written, *no QuickBooks version provides reports with even simple totals for the quantity of items bought or sold!* But all is not lost...

Quantity information the hard way: using a calculator

One way to glean quantity information from QuickBooks reports is by totaling the Qty (quantity) column manually. Here's how:

1. **Open a detail report in QuickBooks.**

   Detail reports list individual transactions and optionally can display the Qty (quantity) column. One example is the Profit and Loss Detail report, which you may open by choosing Reports|Profit & Loss|Detail. (In older QuickBooks versions, choose Reports|Profit & Loss|Itemized.)

2. **Customize the report to include the Qty column.**

   Click on the Customize button in the report window to open the Customize Report dialog. Select the Qty column in the Columns box of the dialog. Then close the dialog by clicking the OK button.

3. **(Optional) Print the report.**

4. **Use a calculator to manually add up numbers in the Qty column.**

   You may total the quantities for individual accounts, etc.

Enhanced quantity reporting with ManagePLUS

The ManagePLUS add-on significantly enhances quantity reporting from QuickBooks accounting records. ManagePLUS reports include quantity totals plus quantity-related statistics such as average price received or paid per bushel, per head, per hundredweight, etc. Here's part of a ManagePLUS report showing quantity totals for a grain sales account, with breakdowns by class and subclass:

Profit & Loss Statement

Includes transactions dated: 4/30/98 - 4/30/99

Printed: 4/30/99

Ordinary Income/Expense

Income
SALES
GRAIN (bu.)

1997 Beans	2,300.00 bu.	$5.46 /bu.	$12,555.00	22	% of BEANS			0.00
1998 Beans	7,900.00 bu.	$5.52 /bu.	$43,576.99	78	% of BEANS	39.50 bu./acre = $217.88 / acre on	200.00 acre	
BEANS	10,200.00 bu.	$5.50 /bu.	$56,131.99					0.00
1997 Corn	30,350.00 bu.	$2.50 /bu.	$75,875.00	43	% of CORN			0.00
1998 Corn	40,450.00 bu.	$2.51 /bu.	$101,580.00	57	% of CORN	161.80 bu./acre = $406.32 / acre on	250.00 acre	
CORN	70,800.00 bu.	$2.51 /bu.	$177,455.00					0.00
Total GRAIN (bu.)	81,000.00 bu.	$2.88 /bu.	$233,586.99					

Enhanced quantity reporting with QuickBooks and a spreadsheet

QuickBooks detail reports are only one step short of providing basic quantity information. They contain the necessary quantity *data*—the quantities entered in individual transactions—but fail to provide totals for them!

So one way to get quantity totals from a QuickBooks detail report is to export it to a file, then import it into a spreadsheet program and add formulas there to total the quantities. This approach has some disadvantages though:

♦ It requires some experience with spreadsheets.

♦ It's difficult to automate adding the spreadsheet formulas, because the exported reports will be arranged a bit differently each time.

♦ You'll have to repeat the process each time you want a report with quantity totals—QuickBooks has no way to save the spreadsheet formulas.

Beginning with **QuickBooks Pro 99,** the Pro versions of QuickBooks **can export any report *directly* into a Microsoft Excel spreadsheet** without the manual steps described below. Spreadsheet formulas, however, must still be added manually.

You'll find detailed steps for exporting a QuickBooks report to a disk file in chapter 12, Essential QuickBooks Reports, but here are the general steps for getting quantity data exported from a QuickBooks report and imported into a spreadsheet program:

1. **Open the desired QuickBooks report window.**

   You must open a *detail* report, because only QuickBooks detail reports contain the necessary quantity data. (Detail reports are those which list individual transactions.)

To open the Profit and Loss Detail report choose Reports | Profit and Loss | Detail (or in older QuickBooks versions, Reports | Profit and Loss | Itemized).

2. **Customize the report to include the Qty column.**

The Qty (quantity) column contains the individual transaction quantities.

To customize the report, click on the Customize button in the report window. Then in the Customize Report window, be sure the Qty item is selected (has a ✔ to the left of it) in the scrollable Columns box. You may have to scroll the Columns box downward to find Qty, then click on it if there's no ✔ beside it. You may also choose other report customization options, such as the range of transaction dates, before clicking OK to close the Customize Report window.

3. **Export the report by "printing" it to a file on your hard disk.**

Click on the Print button in the report window, then in the Print Reports dialog choose to print the report to a file using either the "Tab-delimited file", "Excel/Lotus 123 spreadsheet", or "123 (.PRN) Disk File" option. (Any of these should import into most spreadsheet programs.)

Click on the Print or OK button in the Print Reports dialog, then supply a file name to "print" the report to when prompted for one.

4. **Open your spreadsheet program.**

A spreadsheet program is a separate program, not part of QuickBooks. Many computers come with a spreadsheet program already installed. If you do not have one, you may purchase one from almost any computer dealer.

5. **Issue the necessary commands to import the file (the one you exported from QuickBooks) into the spreadsheet.**

The commands will vary greatly depending on which spreadsheet program you have. For information, look up the topics such as "importing" or "files" in the spreadsheet program's Help system.

6. **Add formulas to the spreadsheet to total the Qty column.**

If you aren't familiar with this, look up "formulas" or "SUM" or "@SUM" in the spreadsheet program's Help system.

7. **Print a copy of the spreadsheet.**

This report should be similar to the QuickBooks report, but with quantities added.

Entering Two Quantities in Income and Expense Transactions (Requires ManagePLUS)

Problem

I want my reports to show livestock purchase and sale weights and the number of head bought or sold. Is there a way to enter two quantities in QuickBooks transactions?

Solution

By itself, QuickBooks only allows entering one quantity per transaction, and only on those forms which allow using items. However, the ManagePLUS add-on supports entering two quantities, and ManagePLUS reports provides totals for both quantities, plus additional statistics whenever two quantities are present, such as average weight per head.

Discussion

ManagePLUS allows entering one or two quantities in the Memo or Description field of any income or expense transaction, including transactions entered on QuickBooks forms which don't support using items, such as deposits and the General Journal.

How to Enter Two Quantities in Transactions

The previous section of this chapter described how to enter quantity information in the Memo or Description field of a transaction. Entering two quantities is about the same except for two additional rules: (1) you must separate the two quantity entries from each other with either an equal sign (=) or a colon (:), and (2) the quantities must always be entered in the same order. For example, if you enter normally enter the weight and a head count in livestock transactions, you must consistently enter the weight first and the head count second (or vice versa); otherwise, the quantities will be totaled incorrectly in reports.

Memo and Description field quantities can also be entered as **mathematical formulas** if you use ManagePLUS, as described on page 181.

Here is a deposit which records the income from selling 44 steers weighing 29,480 pounds and 40 heifers weighing 24,560 pounds:

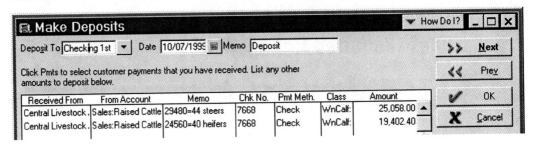

And this is how ManagePLUS would show the transaction information in a report. (Only the two deposit lines shown above are totaled in this report fragment—your reports may typically involve more transactions.)

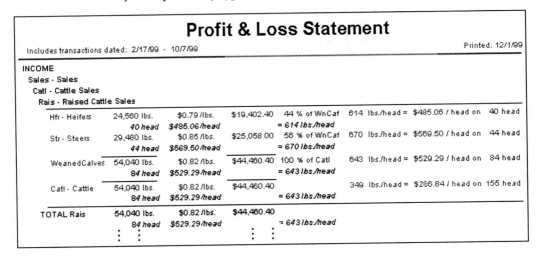

Note that this report provides additional statistics beyond just totaling the quantities. It includes average price received per head and per pound, average selling weight, and class-by-class breakdowns, all calculated automatically by ManagePLUS.

With ManagePLUS you can also enter two quantities on those Quick-Books forms which support using items: the Checks, Bills, Cash Sales, and Invoice forms. The ManagePLUS Help system tells how.

180

Entering Quantities as Mathematical Formulas

Problem *Some of the quantities I enter have to be converted from one unit of measure to another, like converting pounds on a weight ticket to bushels of grain, or converting a count such as the number of bales of hay to tonnage. Is there a way to enter quantities that will let the computer do the math?*

Solution QuickBooks has a built-in calculator feature called Quick-Math, which you can use for calculations in any numeric field. However, only the calculation result gets stored in the transaction. The ManagePLUS add-on lets you enter quantities as formulas in any transaction's Memo or Description field, and defers calculating the result until reports are prepared. This approach preserves the formula details as part of the transaction, so you may review them at any time.

Discussion Don't let the phrase "mathematical formulas" scare you. QuickBooks' QuickMath calculator is as simple as using any other calculator, and entering formulas for processing by ManagePLUS is just as easy. At worst, entering complicated formulas is not much different from the high school algebra we all have forgotten. (Ahem....)

Using the QuickMath Calculator

There's no reason to spend much time or effort describing anything as simple and easy to use as QuickMath, so here are just the important points. QuickMath:

♦ Can be invoked in any numeric field, by pressing any of several special keys.

The calculator will pop up in its own small window when you press the equal sign (=) key. Or if you've already typed a number in the field you can invoke the calculator by pressing any of these keys: + - * / =.

♦ Works like a typical handheld calculator.

Just type a number, followed by one of the arithmetic operation keys (+ - * / =), then another number. You may chain together as many calculations as you like. Here's an example of calculating the number of bushels in a load of corn based on the truck's loaded and empty weights, assuming corn weighs 56 pounds per bushel:

```
      48000
  -   17350
  /      56
  =  547.32143
     547.32143
```

When you're done with a calculation, press the Enter key to have the result stored in the underlying field.

◆ Is only available in numeric fields.

But it never hurts to try invoking the calculator in any field—just type one of the keys mentioned above. If nothing happens you'll know the calculator isn't available in that field.

A numeric field is one where you can enter numbers but not other characters. Fields that handle quantities or dollar amounts are numeric fields. Just because a field *normally* handles numbers though, does not mean it's numeric. For instance, a zip code field is not numeric, because it allows entering letters of the alphabet as well as numbers (some non-U.S. postal codes contain alphabetic characters).

If you want to know more about the QuickMath calculator, consult the Quick-Books Help system—but there's not much more to know!

Entering Quantities as Formulas for Calculation by ManagePLUS

As described earlier in this chapter, the ManagePLUS add-on supports entering quantity information in the Memo or Description field of income and expense transactions. The quantity information may be entered either as numbers or as mathematical formulas. Here are the important points about entering quantities as mathematical formulas in ManagePLUS:

◆ Only works for quantities entered in Memo or Description fields of transaction detail lines, not in Qty (quantity) fields.

Formula results are not calculated until a copy of your QuickBooks transactions are imported into ManagePLUS. The benefit of this is that the formula details always remain a part of the transaction. You'll always be able to see what numbers were used to calculate the quantity, and even change them if you want. The downside is that you don't see the calculation result until reports are prepared in ManagePLUS.

◆ ManagePLUS formulas are more similar to using a spreadsheet than to using a calculator.

When using a calculator you enter numbers individually, one at a time, and the calculator displays a subtotal after each step. In a spreadsheet you enter the entire formula before anything is calculated. That's the way formulas are entered in a Memo or Description field for later calculation by Manage-PLUS.

Also as in spreadsheet formulas, a wider array of arithmetic operators are available with ManagePLUS formulas. For example, you can use parentheses to control the calculation order. Here are some examples of Memo-field quantities entered as formulas in a deposit:

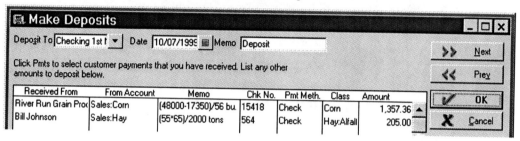

The first detail line of this example calculates the number of bushels in a load of corn based on the truck's loaded and empty weights (48,000 and 17,350 pounds, respectively), assuming corn weighs 56 pounds per bushel. The second detail line estimates the tons of hay on a pickup truck load that was sold by the bale, not weighed. The calculation is based on a sale of 55 bales at an estimated average weight of 65 pounds per bale, divided by 2000 to convert from pounds to tons.

The results of these formulas—547.32 bushels and 1.625 tons—will not be displayed until a report is prepared with ManagePLUS.

Tracking Weighted Average Grade or Quality Information (Requires ManagePLUS)

Problem

I'd like a way to keep track of the moisture content of every load of grain I sell. That would let me figure the weighted average moisture of sales made at harvest versus sales made later. By using classes I could even figure weighted average moisture for different farms or fields, bins, or crop varieties. Can I do this with QuickBooks?

Solution

Whether you are buying or selling something the problem is the same: you need a way to enter grade or quality information in transactions, as a percentage or decimal amount, and get reports which calculate weighted averages based on that information. The ManagePLUS add-on supports entering this kind of information and also provides weighted average calculations on reports.

Discussion

As described in the previous section of this chapter, the ManagePLUS add-on supports entering two quantities in any income or expense transaction. A variation of this capability is entering a grade or quality factor as the second number, instead of a quantity. ManagePLUS uses this information to calculate weighted averages on reports.

Having weighted average information is useful in many types of transactions, whether you are selling grain, hay, or seed, purchasing livestock feed, etc. Examples include weighted average grain moisture, oil content, protein content, Relative Feed Value (RFV—a measure of forage digestibility), seed clean-out percentage, and many others.

What's a Weighted Average?

Suppose you have 10,000 bushels of 10% moisture wheat in a bin and a 500 bushel truck load of 20% moisture wheat. After emptying the truck load into the bin, what is the weighted average moisture content of wheat in the bin?

To calculate the weighted average you multiply each quantity of wheat by its moisture content, add the results together, then divide by the total number of bushels:

$$\frac{(10000 * 10\%) + (500 * 20\%) \text{ moisture}}{10000 + 500 \text{ bushels}} = \frac{1000 + 100 \text{ moisture}}{10500 \text{ bushels}}$$

$$= \frac{1100 \text{ moisture}}{10500 \text{ bushels}} = 10.48\% \text{ weighted average moisture}$$

How to Store Grade or Quality Information in the Memo/Description Field

ManagePLUS supports entering two numbers in the Memo or Description field of deposits and the General Journal, and also on the Expenses tab of the Checks and Bills forms. The first number must always be a quantity, such as the weight or number of bushels bought or sold. The second number may represent another quantity...or a grade or quality factor such as moisture percentage.

An earlier section of this chapter told how to enter numbers in the Memo field of a transaction, in a way that would allow ManagePLUS to extract and use them. The basics are: (1) numbers representing quantities or weighted averages must be entered at the beginning the Memo, (2) only digits and decimal points should be entered, (3) when two numbers are entered they must be separated by either an equal sign (=) or a colon (:), and (4) the numbers must be separated from other Memo text by at least one blank space.

Here is an example of how weighted average information can be added to a transaction. This is a deposit of income from selling 600 bushels of 14.5% moisture corn:

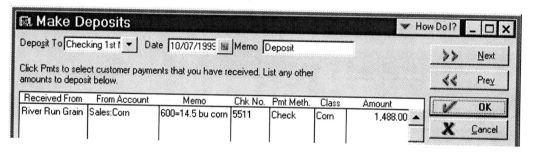

ManagePLUS extracts the Memo field information when it prepares reports, and automatically calculates the weighted average moisture content for all corn sales included in the report. Here's part of a ManagePLUS detail report (it includes a detailed transaction listing) which calculates the weighted average moisture content of fourteen corn sales transactions (the weighted average is toward the bottom right side of the report):

185

Profit & Loss Statement

Includes transactions dated: 1/1/98 - 12/31/98

Printed: 1/5/99

Ordinary Income/Expense
 Income
  Grain
   Corn (bu.)

501.23=16.7 % moisture corn	501.23	River Run Grain Coop.	$1,268.11	Cash Sale	6	1/27/98 Corn:1997:Smith
524=15.7 % moisture corn	524.00	River Run Grain Coop.	$1,400.51	Cash Sale	1	2/28/98 Corn:1997:Smith
548=15.3 % moisture corn	548.00	River Run Grain Coop.	$1,380.96	Cash Sale	7	3/27/98 Corn:1997:Smith
537=15.8 % moisture corn	537.00	River Run Grain Coop.	$1,363.98	Cash Sale	8	4/27/98 Corn:1997:Jones
500=14.2 % moisture corn	500.00	River Run Grain Coop.	$1,275.00	Cash Sale	9	5/27/98 Corn:1997:Jones
513=15.6 % moisture corn	513.00	River Run Grain Coop.	$1,323.54	Cash Sale	10	6/27/98 Corn:1997:Smith
511.41=14.6 % moisture corn	511.41	River Run Grain Coop.	$1,314.32	Cash Sale	11	7/27/98 Corn:1997:Smith
544=14.9 % moisture corn	544.00	River Run Grain Coop.	$1,414.40	Cash Sale	12	8/27/98 Corn:1997:Smith
561=14.6 % moisture corn	561.00	River Run Grain Coop.	$1,469.82	Cash Sale	12	8/27/98 Corn:1997:Smith
545=14.7 % moistures corn	545.00	River Run Grain Coop.	$1,427.90	Cash Sale	12	8/27/98 Corn:1997:Smith
553=22.1 % moisture corn	553.00	River Run Grain Coop.	$1,404.62	Cash Sale	13	9/27/98 Corn:1998:Jones
503=19.9 % moisture corn	503.00	River Run Grain Coop.	$1,252.47	Cash Sale	14	10/27/98 Corn:1998:Jones
527.34=16.4 % moisture corn	527.34	River Run Grain Coop.	$1,265.62	Cash Sale	15	11/27/98 Corn:1998:Jones
552=15.3 % moisture corn	552.00	River Run Grain Coop.	$1,407.60	Cash Sale	16	12/27/98 Corn:1998:Jones

Total Corn (bu.)	7,419.98 bu.	$2.56 /bu.	$18,968.85	*Weighted avg. % = 16.1279*

How to store grade or quality information in a Cash Sales or Invoice form

When using ManagePLUS with QuickBooks forms that have a Qty (quantity) field, an actual quantity must be entered in the Qty field. To add grade or quality information to a transaction, place it in the Memo or Description field.

Here is the same corn sale transaction as described earlier—600 bushels of corn at 14.5% moisture—but entered in the Cash Sales form instead of as a deposit. (It would be entered on an Invoice form in the same way.)

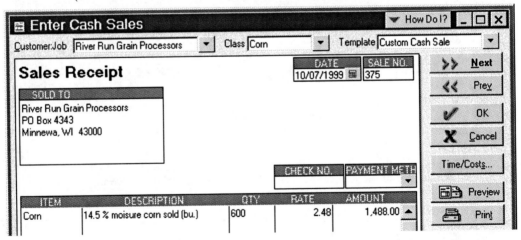

This transaction entry would produce in the same quantity and weighted average information on ManagePLUS reports as the deposit shown earlier.

CHAPTER 8

Special Income, Expense, and Miscellaneous Examples

This chapter covers miscellaneous topics related to income and expense transactions.

Handling Farm Production as Inventory: a Grain Inventory Example

Problem

I understand QuickBooks' inventory system has shortcomings that prevent it from working well with most farm inventories. But is there a way I can make it work for the inventories of things I produce, like grain?

Solution

If you set up Inventory Part type items in a non-standard way, it is possible to use QuickBooks' inventory system to keep track of farm production.

Discussion

To understand why a special approach is needed for farm inventories, a bit of background is in order. QuickBooks takes an accrual accounting approach to the cost of inventoried items, which means the cost of the items is deducted as a business expense when the inventory is sold. Most farm businesses use cash-basis accounting, in which production costs are deducted when production inputs are purchased. Allowing the inventory system to also deduct production costs when inventory is sold would double-count the production costs!

You might think this problem could be avoided by entering farm production into inventory at a cost of zero. True, that would solve the problem of double-counting production costs, but it would raise another problem: the inventory would also be listed at a value of zero on balance sheet reports!

For a brief discussion of QuickBooks' shortcomings for farm inventories, see the shaded box titled "QuickBooks' Farm Inventory Problems...", on page 280.

This section describes an easy work-around which allows QuickBooks to handle inventories of farm *production* (but not farm inputs) moderately well—it's at least an adequate way to get the value of production inventories into the farm balance sheet. The only negative part of the work-around is that QuickBooks' inventory adjustment window must be used for getting production into inventory. But even that isn't really difficult, once you understand how inventory adjustments are supposed to work.

Here are the main points of the work-around technique:

♦ Set up Inventory Part type items for farm production that is to be handled as inventory. Select Retained Earnings (or any equity account) in the COGS Account field of the item setup dialog. (This will cause cost of goods sold postings for the item to be discarded, so they won't be posted as expense.)

♦ When farm production is harvested, or you want to include things like market livestock or growing crops in a balance sheet, make inventory adjustment entries to add the value of these things to inventory.

♦ To sell items from inventory, use either the Cash Sales form or the Invoice form. These are the only QuickBooks forms which deduct quantities from inventory when you enter a sale.

♦ You may get reports of inventory value and quantity on hand at any time.

♦ Before preparing a balance sheet, adjust the quantity and value of inventory items to match the physical count of inventory on hand.

The examples in this section describe accounts, items, and activities for managing a Corn inventory item, but the same techniques apply to *any* type of farm production you want to handle as inventory in QuickBooks—grain, hay, produce, growing crops, raised market livestock, etc. Note however, that livestock purchased for resale and other resale items should be handled differently, as described in the topic beginning on page 199.

How to Set Up the Accounts You Will Need

Three accounts are required for setting up any Inventory Part type item. Here are the steps for setting up accounts necessary for a Corn item (which will be added later):

1. **Open the Chart of Accounts window.**

Choosing Lists|Chart of Accounts is one way to do this.

2. **Open the New Account dialog.**

Either type Ctrl-N, or choose Edit|New Account, or click on the Account button in the lower part of the Chart of Accounts window and then select New from the pop-up menu.

3. **Set up an income account for corn sales, if you don't already have one.**

When the Corn item is used in a sales transaction, the revenue will be posted to this account. For this example, we'll set up a Sales:Grain:Corn income account (Corn is a subaccount of Grain, which is a subaccount of Sales):

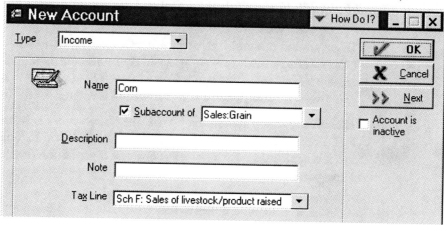

4. **Set up an inventory asset account for corn, if you don't already have one.**

An asset account is necessary for representing the corn's value on the balance sheet. For this example we'll set up a Grain Inventory:Corn account. Like most inventory asset accounts, this one is created as an Other Current Asset type account:

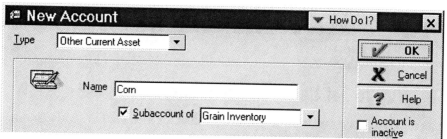

By the way, it would have been possible to use Grain Inventory as the asset account without setting up a separate subaccount for corn. The advantage of having a Corn subaccount is that it will cause the corn inventory to be listed separately on balance sheet reports. (The number of inventory accounts you set up determines how much detail will appear on balance sheets.)

5. **Decide which account you'll use as a cost of goods account for the Corn item.**

This is a key part of the work-around!

Normally an account named Cost of Goods Sold is selected in the COGS Account field (of the New Item dialog) when setting up an Inventory Part item. Whenever the item is used in a sales transaction, its cost will be posted to that account. The problem is, Cost of Goods Sold is *deducted* on profit and loss reports, and that causes the double-counting of expenses mentioned earlier.

To get around this problem, don't select Cost of Goods Sold in the item's COGS Account field. Instead, select an equity account. The goal is to use an account which causes the item's Cost of Goods Sold posting to be automatically "thrown away" when a sales transaction is entered for the item.

Though any equity account may work, the easiest choice is Retained Earnings. Every QuickBooks chart of accounts has a Retained Earnings account, and it is appropriate for all types of business organization (sole proprietorships, partnerships, and corporations).

Retained Earnings is a special account. Retained Earnings amounts on balance sheets are *calculated*—QuickBooks does not maintain a balance for Retained Earnings, as you can see if you look at the Chart of Accounts window. That often makes Retained Earnings useful when you want to discard the offsetting part of a transaction.

Opening Bal Equity (Opening Balance Equity) is another good account choice, as it is also present in every QuickBooks chart of accounts unless it has been renamed.

6. **Click OK to close the New Account dialog if you haven't already done so.**

How to Set Up Inventory Items

Set up one Inventory Part type item for each kind of production inventory you want to track in QuickBooks. For example, you might have items named Corn, Soybeans, Wheat, and Market Hogs. Here are the steps for setting up a Corn inventory item:

1. **Open the Items List window.**

Choose Lists|Items.

2. **Open the New Item dialog.**

   Either type Ctrl-N, or choose Edit|New Item, or click on the Item button in the lower part of the Items List window and then select New from the pop-up menu.

3. **Set up the new item.**

   Here is an example of how to set up a Corn item, followed by a discussion of some of the important fields:

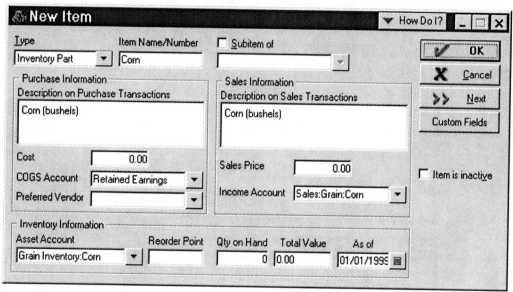

Item Name

Use any item name that will remind you of the item's purpose. "Corn" is adequate for this item, because there's no confusion about how the item will be used.

If a question could arise about which unit of measure is appropriate for an item, find a way to indicate the unit of measure in the item's name. In some areas of the U.S., for example, grain sorghum is sold both in bushels and in hundredweights. If you prefer to keep inventory quantities in bushels, you might name a grain sorghum item "Sorghum-bu." to remind you that transactions for the item need to be entered in bushel units.

Description

Including the unit of measure in the Description fields is another reminder of the proper unit of measure for the item.

COGS Account

Very important! For this special item setup to work as described, you need to select an equity account, such as Retained Earnings, as the COGS (Cost of Goods Sold) Account.

Income Account

Select the account you want to use for recording income from the sale of the item.

Asset Account

The account you select in this field will represent the item's inventory value on balance sheets.

4. **Click OK to close the New Item dialog.**

How to Get Production into Inventory

At harvest time (also, prior to preparing a balance sheet) you need to get the amount of production on hand entered as inventory. The simplest way to do this is by making an inventory adjustment in the Adjust Quantity/Value on Hand window. Here are the steps involved:

1. **Open the Adjust Quantity/Value on Hand window.**

   Choose Activities|Inventory|Adjust Qty/Value on Hand.

   Here's how the window will look after completing the steps that follow (it will look different when first opened):

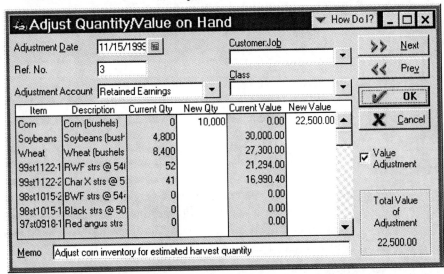

2. **Select the Value Adjustment check box (it's on the right side of the window, below the buttons).**

This causes a New Value column to be added to the window, as the illustration shows. (This step is only necessary if you are adjusting *both* the quantity and the value of inventory items. If you're only adjusting quantities it is not necessary.)

3. **Select an account in the Adjustment Account field.**

    The account you select here will be used as the offsetting account for the adjustment entry. Which account should you use? The appropriate account depends on the type of inventory adjustment being made. For harvested grain the inventory increase means owner's equity has increased also, so selecting any equity account will do. Retained Earnings was selected in the example.

Corporations should *never* post inventory adjustments to accounts which represent stockholder equity. If your accounting method allows the inventory adjustments described here, always post them to some other equity account such as Retained Earnings. Consult an accounting professional about permissible inventory value adjustments in your particular farm business situation.

4. **Enter the quantity of corn on hand in the New Qty column (of the Corn row).**

5. **Enter the value of the corn in the New Value column (of the Corn row).**

    In this example the value is $22,500, which means the corn is worth $2.25 per bushel. (Note that the New Value amount is the value of the *entire* New Qty quantity, not of one unit.)

    By the way, it would have been possible to adjust the quantities and values of other inventory items in this same entry. To keep the example simple, the Corn inventory is the only one adjusted here.

6. **Click on the OK button to save the adjustment entry and close the window.**

    If you want to verify that the adjustment was made correctly, open the Chart of Accounts window (Lists|Chart of Accounts) and find the Grain Inventory:Corn account. The account balance should match the New Value amount entered in step 5 (which was $22,500 in this example).

With the harvested corn entered as inventory, balance sheet reports will now include the corn's value, and corn sales transactions will decrease both the quantity on hand and the remaining value of the inventory.

How to Enter Sales of Inventory Items

Entering sales of Inventory Part items is the same as for other types of items. The only difference is in the results (how account balances are affected), and QuickBooks takes care of all that. See chapter 6 for examples of using the Cash Sales form to enter sales involving items.

Sales of inventory items *must* be entered on either the Cash Sales form or the Invoice form. No other QuickBooks sales form is able to adjust inventory quantities and value for the amounts sold.

How to Get Reports of Inventory on Hand

QuickBooks has several reports you may use, to get a listing of the quantities and/or value of inventory items on hand. One popular choice is the Inventory Valuation Summary (Reports|Inventory Reports|Valuation Summary), which lists both the quantity and the value of each inventory item:

John Doe Farms
Inventory Valuation Summary
As of December 31, 1999

	Item Description	On Hand	Avg Cost	Asset Value	% of Tot Asset
Corn	Corn (bushels)	10,000	2.25	22,500.00	19.1%
Soybeans	Soybeans (bushels)	4,800	6.25	30,000.00	25.4%
Wheat	Wheat (bushels)	8,400	3.25	27,300.00	23.1%
99st1122-1	RWF strs @ 546 lb...	52	409.50	21,294.00	18.0%
99st1122-2	Char X strs @ 592...	41	414.40	16,990.40	14.4%
ZZResaleLvstk	(Archival) Resale ...	0	0.00	0.00	0.0%
TOTAL		**23,293**		**118,084.40**	**100.0%**

About Adjusting Inventory Values...

At various times and for various reasons you will need to adjust inventory quantities and/or value. Here are some of them (adjustment examples are provided farther below):

♦ **When inventory is produced.**

Getting production into inventory was discussed earlier.

♦ **Before preparing a balance sheet.**

If you prepare market value balance sheet reports, you will need to update inventories to current quantities and estimated market value.

♦ **When inventory is consumed on the farm.**

When grain or hay are fed to livestock, for example, they will be used up without being deducted from QuickBooks' inventory count. The inventory count will then need to be adjusted to reflect the actual quantity on hand. Optionally, the cost of inventory consumed by specific enterprises may be charged to those enterprises (classes).

♦ **When inventory is destroyed or lost.**

Besides sales or consumption of inventory, the inventory quantities in Quick-Books may differ from actual amounts on hand due to spoilage, theft and other casualty losses, and mis-estimation of the quantities originally entered into inventory.

Taking a physical inventory count...

The only way to verify and correct QuickBooks' inventory counts is to compare them against the actual physical quantities on hand. To make the job of counting inventory easier, QuickBooks has a physical inventory worksheet you can use. It lists all the inventory items and has space for writing in a quantity for each one, which helps you collect all the necessary information before making an inventory adjustment entry.

To get a copy of the worksheet, choose Reports|Inventory Reports|Physical Inventory Worksheet, then click on the Print button in the report window to print out a copy of it. Here's an example of the worksheet:

John Doe Farms

Physical Inventory Worksheet

All Transactions

	◊ Item Description ◊	◊ On Hand	◊ Physical Count ◊
Corn	Corn (bushels)	10,000	_____
Soybeans	Soybeans (bushels)	4,800	_____
Wheat	Wheat (bushels)	8,400	_____
99st1122-1	RWF strs @ 546 lb...	52	_____
99st1122-2	Char X strs @ 592...	41	_____
ZZResaleLvstk	(Archival) Resale ...	0	_____

Adjusting Inventories to Market Value (for Balance Sheet Preparation)

Before preparing a market value balance sheet you will normally need to adjust inventory quantities and value to estimated current market value. Here's an entry which adjusts the value of the Corn inventory, assuming corn is now worth $2.40 per bushel:

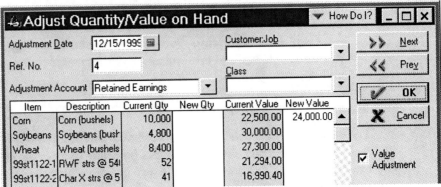

Note that an equity account (Retained Earnings) is again used as the offsetting account for the adjustment. In cash-basis accounting the increase in inventory value is not considered income, just a change in owner's equity.

Inventory value adjustments work for including the investment in a growing crop or market livestock in the farm's balance sheet.

Just set up Inventory Part items for the inventories you want shown on the balance sheet, then assign them a value (and a quantity of at least "1") with adjusting entries like those shown here.

Adjusting Inventories for Amounts Used on the Farm (as Livestock Feed, etc.)

When inventory is consumed on the farm—as when grain or hay are fed to livestock—QuickBooks' inventory count needs to be decreased to match the actual remaining quantity on hand.

If you only want to correct the inventory count and value, an adjustment entry like the one shown above will work fine. Adjust the quantity and value of the items that have been consumed, and use Retained Earnings or some other equity account as the offsetting account.

196

If you keep detailed enterprise records though, you may want to charge a specific enterprise (class) with the value of inventory consumed. You can do this by posting separate adjustment entries to different accounts or classes. For an example, see the shaded box titled "Charging corn consumption to the hog enterprise", on page 197.

Adjusting for Losses from Inventory

Other reasons why QuickBooks' inventory count may differ from the actual quantity on hand are spoilage; theft, fire, flood, and other casualty losses; and mis-estimation of the original quantity placed in inventory.

For cash-basis record keepers, adjustments for these inventory changes are the same as for others: just update the quantity and value, and use an equity account as the offsetting account for the adjustment. (Cash-basis record keepers *may not* deduct casualty losses of inventory because the costs of producing the inventory have already been deducted as expense.)

For accrual record keepers, casualty losses of inventory *are* deductible in certain circumstances. Consult a tax professional for details.

Advanced technique:
Charging corn consumption to the hog enterprise

If inventoried farm production is consumed on the farm, it is possible to charge a specific enterprise class with the value of the consumed inventory. For example, if Corn is an Inventory Part type item you could charge a Hogs class with the value of corn fed to hogs.

Note: with the technique described here it is <u>not</u> possible to credit a Corn class with income, only to charge the consuming enterprise (Hogs) with the value of the corn that was fed.

The first thing you need is one or more special-purpose accounts to record the transferred expense. Accounts of the Expense type shouldn't be used because they appear in the main part of profit and loss reports and would cause expenses to be double-counted. (The value of the fed corn would be listed as an expense, and so would the costs of producing the corn.)

A better choice would be to set up accounts of the Other Expense type. While it is true that amounts posted to Other Expense accounts are also deducted on profit and loss reports, they are listed in a separate section at the bottom of the report—where their contribution to Net Income may be ignored. Here is an example New Account entry for a feed expense transfer account:

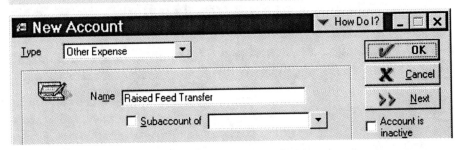

Do not assign a tax line to expense transfer accounts, or expenses will be double-counted on income tax reports.

To use this account, select it as the offsetting account in an inventory adjustment entry. If QuickBooks shows a 10,000 bushel inventory for the Corn item but you estimate only 4,000 bushels remains on hand, you may assume the other 6,000 bushels were fed to hogs. You could transfer the value of the fed corn to Raised Feed Transfer and to the Hogs class with this entry:

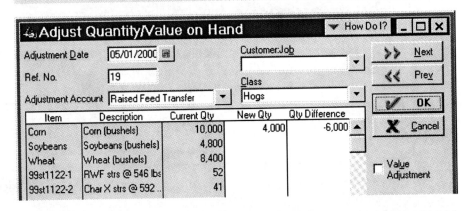

This adjustment transfers $14,400 (6,000 bushels x the current inventory value of $2.40 per bushel) to the Raised Feed Transfer account, and also assigns it to the Hogs class. (This amount will appear in any report of expenses for the Hogs class.)

Keeping Track of Livestock Purchased for Resale

Problem
How should I keep track of livestock purchased for resale, for proper income tax records?

Solution
There are a couple approaches you may use. Neither are automatic, nor simple. Both require a bit more effort than ought to be necessary, but they work!

Discussion
The two main goals of tracking livestock purchased for resale are: (1) to be able to identify their purchase cost, as required for reporting Federal income taxes, and (2) to include their value on the farm balance sheet. Unless you don't prepare farm balance sheets with QuickBooks, any method you use to keep track of livestock purchased for resale should achieve both goals.

Two methods for handling resale livestock are described in this section. The first uses account entries to track the livestock's purchase cost, and to transfer the purchase cost to a cost accumulation account when livestock are sold.

The second method uses QuickBooks inventory features. As described elsewhere, QuickBooks' inventory features only really work well for retail items—things that were purchased by the business to sell at retail. Resale livestock don't exactly fit that mold, but they are similar enough that, with care, you can use QuickBooks' inventory features to properly track them.

Livestock purchased for resale is the IRS term for what producers typically call feeder livestock. These are livestock purchased for feeding—not breeding—purposes. They are kept for a period of time, usually less than one year, before they are sold.

Method #1: Handling Resale Livestock Inventories Manually

You could say this is a "textbook" approach to accounting for resale livestock. It doesn't provide any special features or information and requires some manual work on your part, but it gets the job done without being overly complicated. Here are the basics of the process:

- Set up asset accounts to represent the value of livestock purchased for resale.

- When you buy feeder livestock, store their purchase cost in one of these asset accounts.

- When you sell the livestock:

- Look up their original purchase cost (in the asset account's Register window)

- Record the entire sale proceeds as income.

- Transfer the original purchase cost from the asset account to an expense account. (This can usually be done as part of the same entry which records the income.)

Setting Up the Accounts You'll Need

You will need to add several accounts to keep track of resale livestock. Here are the steps for setting the up:

1. **Open the Chart of Accounts window.**

   Choosing Lists|Accounts is one way to do this.

2. **Open the New Account dialog.**

   Either type Ctrl-N, or click on the Account button in the lower part of the Chart of Accounts window and then select New from the pop-up menu.

3. **Set up each account described below, clicking the Next button after each one, and the OK button when you're done with all of them.**

   You need one or more asset accounts to represent the value of the livestock when they are first purchased:

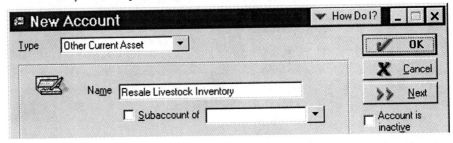

You also need an income account for recording income when you sell the livestock, and an expense account to which the purchase cost of the livestock will be transferred when you sell them. Examples of both accounts are shown below.

Note that a tax line is selected for both accounts, so the income and expense amounts posted to them will be included when you generate an Income Tax Report from QuickBooks.

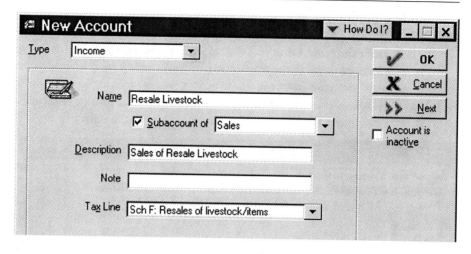

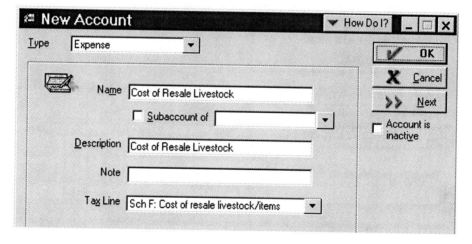

 An alternative account setup is to make both the sales and cost accounts subaccounts of the same income account:

Sales	10,000
Resale Livestock	10,000
Sales	50,000
Cost	40,000

You would then record income in the Sales:Resale Livestock:Sales account, and the livestock's purchase cost in Sales:Resale Livestock:Cost. The balance of the parent account (Resale Livestock) would always be the difference between the Sales and Cost subaccounts' balances.

Entering a Purchase of Resale Livestock

Enter the purchase as you would enter any asset purchase. Just be sure to include plenty of detail in the transaction so if you later look it up to determine the animals' purchase cost you can be certain you are looking at the right transaction. Here's an example check written for two different groups of feeder calves, with extra lines added to store pertinent details with the transaction:

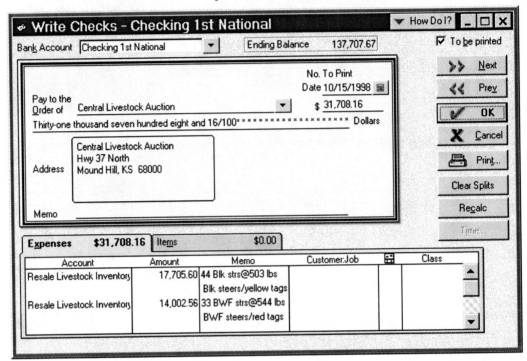

This entry increases the balance in the Resale Livestock Inventory account by $31,708.16.

Entering a Sale of Resale Livestock

There are several things to accomplish when you enter a sale of livestock purchased for resale. As an example, let's assume 20 of the BWF steers are sold—they're the second group listed in the check entry, above. The following two sections show how to (1) look up their purchase cost, then (2) enter income from the sale and transfer the steers' purchase cost to a resale livestock expense account (so it will be included as an expense in income tax reports).

First, look up the purchase cost:

The first thing to do is determine the steers' average per-head purchase cost, which you can do by looking back at the purchase transaction in the Resale Livestock Inventory account:

1. **Open the Chart of Accounts window.**

   Choosing Lists | Chart of Accounts is one way to do this.

2. **Click on the Resale Livestock Inventory account, to highlight it.**

3. **With the account highlighted, open the account's Register window.**

   Either type Ctrl-R, or click on the Account button at the bottom of the Chart of Accounts window and then click on Use Register in the pop-up menu.

   Here is a partial view of the Register window, showing the two transaction lines entered when the cattle were purchased (the check entry shown earlier). The purchase figures for the 33 BWF steers can be used to calculate their average cost: $14,002.56 / 33 head = $424.32 per head.

Resale Livestock Inventory				How Do I?		
Date	Ref	Payee		Decrease	Increase	Balance
	Type	Account	Memo			
10/15/1998		Central Livestock Auction			17,705.60	17,705.60
	CHK	Checking 1st National [split] 44 Blk strs@503 lbs				
10/15/1998		Central Livestock Auction			14,002.56	31,708.16
	CHK	Checking 1st National [split] 33 BWF strs@544 lbs				

4. **(Optional) Close the Register window.**

   You may want to leave it open, to allow referring back to the animals' purchase cost, as you make the other entries.

Second, enter the income and transfer the purchase cost:

Next, enter the income from selling the steers and transfer the purchase cost to an expense account. You may either (1) enter the income first as a cash sale or deposit, then make a General Journal entry to transfer the purchase cost to the expense account, or (2) accomplish both within one deposit entry, as shown here:

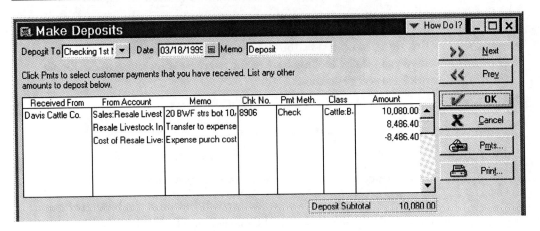

The first line in the detail area of the deposit records the income from selling the 20 steers. The second and third lines accomplish a transfer of the steers' purchase cost (20 head x $424.32 per head = $8486.40) from Resale Livestock Inventory to Cost of Resale Livestock. (The steers' inventory value is transferred to an expense account.) Notice that the third line is *negative*, so the second and third lines cancel each other out, and their net effect on the deposit is zero.

Why is the Cost of Resale Livestock amount entered as a *negative* number?

A positive number in a deposit entry normally records income. To properly record expense in a deposit, a number must be negative.

What if some animals from *both* purchase groups had been sold? The deposit entry would be similar, except that the amount to transfer from Resale Livestock Inventory to the Cost of Resale Livestock account would be based on the number of animals sold from each groups, since their purchase costs were different.

What About Death Losses?

When deaths occur in livestock that were purchased for resale, the purchase cost of the dead animals is deductible from income. You need to transfer the appropriate amount of the purchase cost from the Resale Livestock Inventory account to the Cost of Resale Livestock expense account.

The General Journal is one place where you could enter the transfer. But if you are confused by debits and credits a better choice would be to enter death loss in the Register window of the Resale Livestock Inventory account, as the following steps describe:

Register windows are often easier to use than the General Journal because they use the terms "increase" and "decrease" in place of "debit" and "credit".

1. **Open the Chart of Accounts window.**

   Choosing Lists | Chart of Accounts is one way to do this.

2. **Click on the Resale Livestock Inventory account, to highlight it.**

3. **With the account highlighted, open its Register window.**

   Either type Ctrl-R, or click on the Activities button at the bottom of the Chart of Accounts window and then click on Use Register in the pop-up menu.

4. **Enter the transfer in the register.**

   Here's part of the Register window for the Resale Livestock Inventory account. The last line is the new entry which transfers the purchase cost of one steer to the Cost of Resale Livestock account:

Resale Livestock Inventory

Date	Ref	Payee		Decrease	Increase	Balance
	Type	Account	Memo			
10/15/1998		Central Livestock Auction			17,705.60	17,705.60
	CHK	Checking 1st National [split]	44 Blk strs@503 lbs			
10/15/1998		Central Livestock Auction			14,002.56	31,708.16
	CHK	Checking 1st National [split]	33 BWF strs@544 lbs			
03/18/1999				8,486.40		23,221.76
	DEP	Checking 1st National [split]	Transfer to expense			
03/20/1999				424.32		22,797.44
	GENJRNL	Cost of Resale Livestock	1 steer died			

If you want to keep track of death losses separately, an alternative would be to set up an expense account called Death Losses and post the transfer to it rather than to Cost of Resale Livestock.

Method #2: Handling Resale Livestock Using Inventory Part Type Items

This technique is for users who have at least a basic understanding of the QuickBooks inventory system.

Bending the Rules...

With care, you can "bend" the intended purpose of QuickBooks items to keep track of livestock purchased for resale. Here are the basics—you can fill in the details on your own:

1. **Set up a *new* Inventory Part item type each time a group of livestock is bought for resale, and assign the cost of one animal in the item's Cost field.**

   Considering the same example as earlier in the chapter, you would need to set up one new Inventory Part item for the 44 black steers that were purchased, and another for the 33 steers. Here's how the Inventory Part item setup might look for the lot of 44 black steers that were purchased, plus comments on some of the important fields:

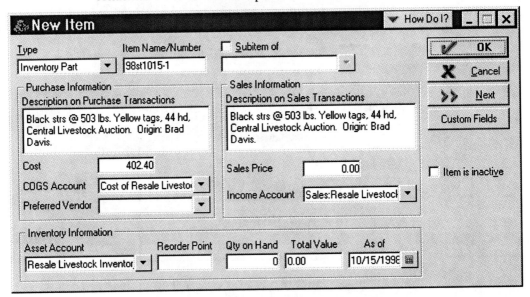

Item Name
You can assign any item name, but since you may be creating a lot of these items (one for *each* group of livestock purchased), it's best to have some kind of an organized plan for naming items—hopefully one that makes them easy to locate on reports and in the Items list. The item name in this example is composed of "98" for the year, "st" for steers, "1015" for the month and day of purchase, and "-1" to separately identify different lots purchased on that date.

Description
This should be adequate to let you precisely identify the group of animals that was purchased.

Cost

Very important! Assign the purchase cost of *one* animal in this field, by dividing the total cost by the number of animals in the group. This will let you use the item to record sales or death loss involving any number of animals from the group.

COGS Account

Select the expense account you use to accumulate the cost of resale livestock for tax purposes. That way, using the item to record a sale of livestock will automatically transfer the purchase cost amount to that expense account.

Income Account

Select the account you use to record income from the sale of resale livestock.

Asset Account

The asset account you select here will reflect the purchase cost of resale livestock on hand, on farm balance sheets.

Another Inventory Part item would be needed for the group of 33 BWF steers:

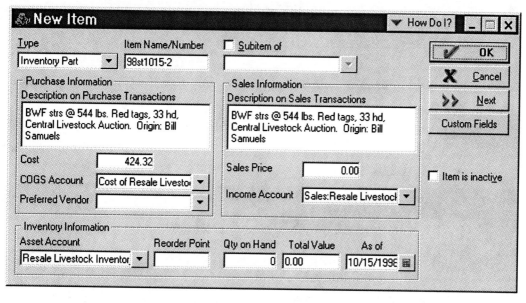

2. **Use the new item(s) to record the purchase.**

Here's the same livestock purchase illustrated in the Checks form on page 202, but this time entered on the Items tab of the Checks form:

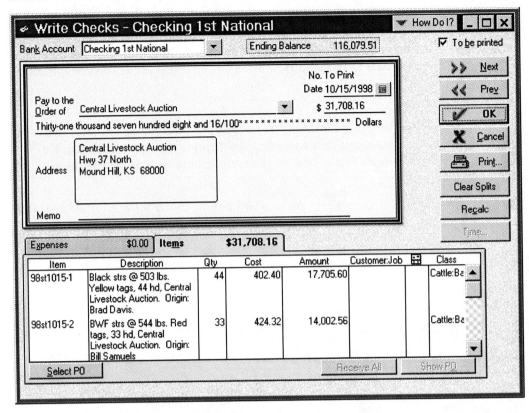

3. **When animals are sold, record the sale on either the Cash Sales or Invoice form, by selecting the appropriate item(s).**

Here's the detail area of a cash sale entry which records the sale of 20 head of the steers:

ITEM	DESCRIPTION	QTY	RATE	AMOUNT
98st1015-2	BWF strs Avg: 715 lbs. Bought 10/15/98 Origin: Bill Samuels.	20	529.11	10,582.20

Because this entry uses Inventory Part items, it accomplishes all of the same accounting jobs as the entire three-line deposit entry on page 204!

4. **If an animal dies, enter a Cash Sale at a selling price of zero.**

This will transfer the cost of the animal to the Cost of Resale Livestock expense account, but records income of zero. Here's the detail area of a Cash Sales entry for death loss of one steer:

ITEM	DESCRIPTION	QTY	RATE	AMOUNT
98st1015-2	BWF strs - 1 died, bought 10/15/98 Origin: Bill Samuels	1	0.00	0.00

5. **Periodically view a report of inventory counts and values.**

This will serve as a check on whether your QuickBooks inventory counts match actual livestock numbers on hand. If not, you may need to either record previously unrecorded death loss, or use QuickBooks' Adjust Inventory Quantity and Value activity (Activities|Inventory|Adjust Qty/Value on Hand) to fix the inventory counts and value.

Comments...and Warnings

◆ The chief advantage of this method comes when recording sales of resale livestock. Entering sales using the Inventory Part items that were created when the animals were purchased will record the sales income and will *automatically* transfer the animals' purchase cost to an expense account.

◆ The disadvantage of this method is that it requires frequent additions to the Items list. You may not run out of space for items any time soon (QuickBooks allows creation of up to 14,500 items), but adding too many items to the Items list will make it difficult to use. To keep the Items list pared down to a reasonable number of items, hide the inactive items so QuickBooks won't display them. See page 271 for details on hiding inactive items.

◆ QuickBooks uses the average costing inventory valuation method. That is why you must create a *new* item for each purchase of resale livestock and *cannot* reuse items set up for this purpose. If you don't set up a new item each time, the purchase cost accumulated in the Cost of Resale Livestock account will be incorrect!

◆ Because Inventory Part type items are used, you can generate QuickBooks' inventory reports to see a count of how many animals are still on hand, from each purchase group.

Entering Milk Sales Income

Problem

How should I enter milk checks so that partial payments, capital retains, and deductions for things like milk hauling are handled properly? And how should I record equity in the dairy cooperative? What about capital (equity) revolvements?

Solution

This section provides examples of milk sales and related transactions.

Discussion

The Cash Sales form is a good place to enter milk sales because it lets you record the quantity of milk sold, plus other necessary parts of the transaction.

Setting Up the Accounts, Classes, and Items You Will Need

Accounts

Several accounts are necessary to keep track of the various parts of a milk sales transaction. Here are accounts used in this section's examples.

Account	Account Type	Comments
Capital Retain Certificates	Other Asset	Represents the value of your equity in the dairy cooperative, resulting from capital retain deductions from your milk checks.
Sales	Income	Parent account of all sales subaccounts.
Sales:Milk	Income	Milk sales (subaccount of Sales).
Advertising	Expense	Advertising expenses, including advertising expense deductions from milk checks
Capital Retain Deductions	Expense	Capital retain deductions from milk checks.
Freight & Hauling	Expense	Freight and hauling expenses, including those deducted from milk checks.

To add these accounts to the Chart of Accounts:

1. **Open the Chart of Accounts window.**

   Choosing Lists|Chart of Accounts is one way to do this.

2. **Open the New Account dialog.**

   Either type Ctrl-N, or click on the Account button in the lower part of the Chart of Accounts window and then select New from the pop-up menu.

3. **Set up the information for each account, clicking the Next button after each one, and the OK button when you're done with all of them.**

   Here are partial views (showing just the important fields) of the New Account dialogs for each of the accounts listed above. Note that tax lines are selected for accounts where they are appropriate, which will include the milk sales income and expenses in Income Tax reports printed from QuickBooks.

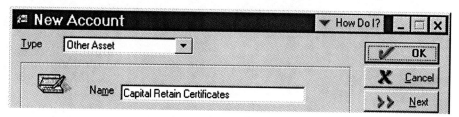

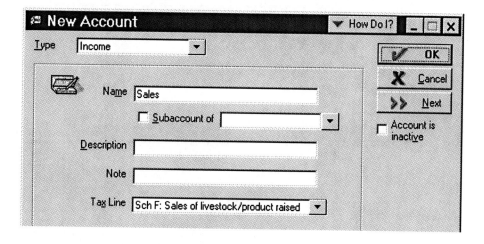

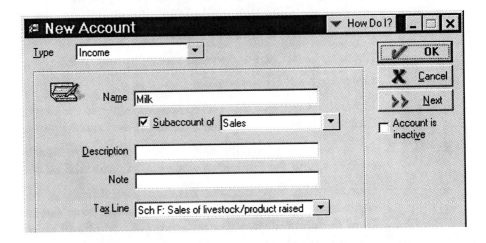

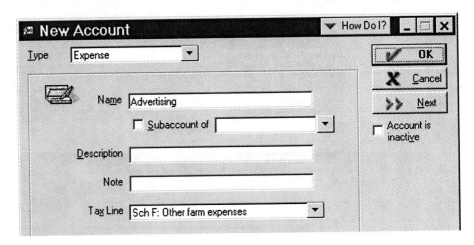

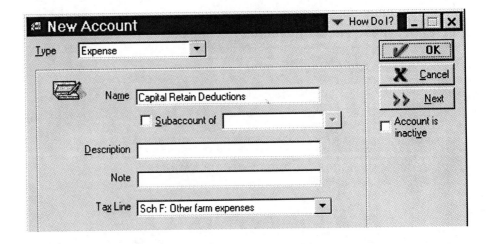

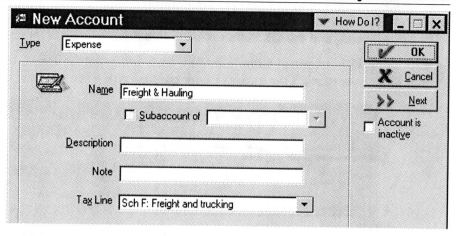

Items

If you enter milk sales on the Cash Sales form, you'll need to set up these items to identify the various parts of the transaction:

Item Name	Type	Description
Milk	Non-inventory Part	Milk sales (parent item).
Milk:Partial	Non-inventory Part	Partial-month milk payment (subitem of Milk).
Milk:Full	Non-inventory Part	Full-month milk payment (subitem of Milk).
CapRetain	Discount	Capital Retain DEDUCTION.
Advert	Discount	Advertising DEDUCTION.
Hauling	Discount	Hauling DEDUCTION.

To add new items to the Items list:

1. **Open the Items list.**

   Choosing Lists|Items is one way to do this.

2. **Open the New Item dialog.**

   Either type Ctrl-N, or click on the Item button in the lower part of the Items List window and then select New from the pop-up menu.

3. Fill in information for the new items, clicking the Next button after each one, and the OK button when you're done with all of them.

Here are partial views of the New Item dialogs for the items listed above. Be sure to select the correct account and item type when you enter them:

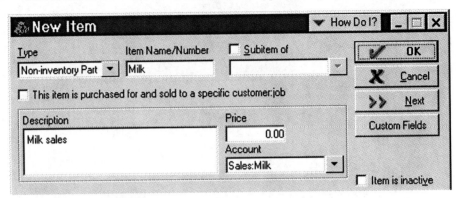

Note that the following two items are *subitems* of the Milk item.

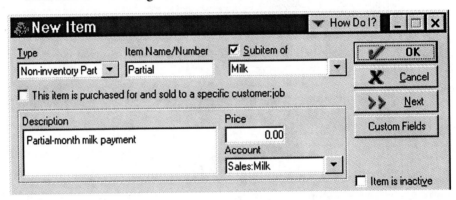

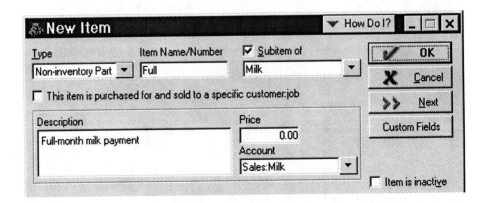

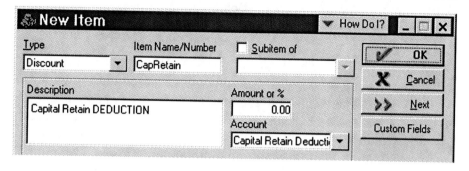

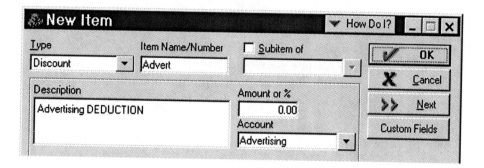

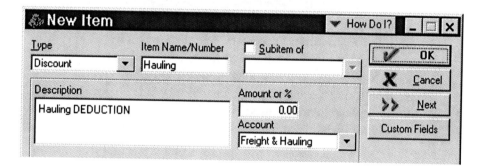

Classes

Using classes isn't a requirement, but this is a good place to mention ideas about using classes in a dairy operation. You could set up just one Dairy class and use it to tag all income and expenses related to the dairy enterprise. Or you could go farther and set up subclasses to identify more specific profit centers within the main Dairy class. Here are some examples:

> Dairy
> Calves
> Milking Herd
> Heifer Development

These subclasses allow gathering specific information about various dairy activities. If you feed calves instead of selling them off of the cow, the Calves subclass would allow tracking feed and other expenses and income from the calves. The Milking Herd subclass is for tracking expenses and income related to the milking herd, including income from milk sales and cull cows. The Heifer Development subclass could track costs of raising replacement heifers, either for your own herd or to sell.

> A **profit center** is any subdivision of the farm business you can identify as having its own inputs and output, and which is operated with the intent of making a profit. Profit centers usually correspond with the things you produce, like corn, cotton, or feeder cattle...or in the accompanying example, with calves, milk production, or heifer development.

How to Enter a Partial Payment for Milk

Some dairy producers receive a milk check at mid month as a partial payment, followed by a check at the end of the month for the full month's sales, less deductions.

If you receive partial payments, enter them as milk sales income on either the Cash Sales form or in a deposit. In either case it's usually best *not* to enter the milk quantity when recording a partial payment. Wait until you receive the full-month check, and enter the entire month's milk sales quantity when you record it. This eliminates confusion over what part of the full-month quantity may have been recorded when the partial payment was entered.

Here's a cash sale entry for a partial milk payment. Notice that the Qty (quantity) column has been left blank—only dollars are recorded.

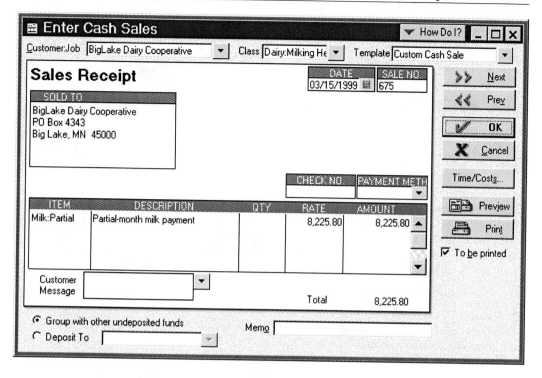

How to Enter a Full Payment for Milk

A milk check is like many other checks you receive: it pays you a net amount left over after some expenses or fees have been deducted from a gross amount. What makes milk checks a bit unique, is deductions for capital retains (member equity contributions to a dairy cooperative) and for partial payments received earlier during the month.

The Cash Sales form is usually the best place to enter milk sales. It lets you use items to identify specific types of income and deductions, and it gives you a place to enter the quantity of milk sold. Once you've created the necessary item definitions, described earlier, entering the check is easy:

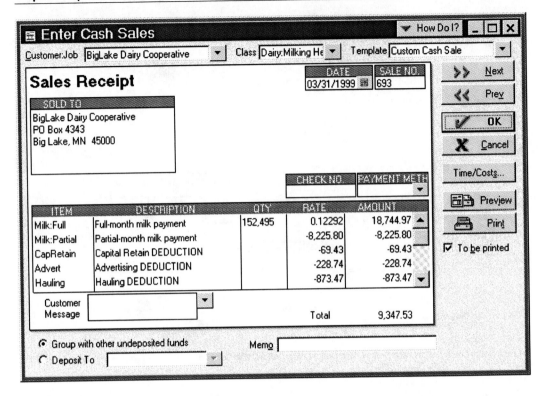

Notes:

- The milk quantity sold during the entire month is entered in the Qty column, because it was not included on the partial-month payment's entry.

- You don't have to fill in the Rate column (price per pound of milk). Just enter the Qty and the Amount, and QuickBooks will calculate the Rate for you.

- The partial payment received earlier in the month (if any) is entered as a deduction—as a negative dollar amount.

- Capital retains are deducted as an expense. The IRS doesn't consider them taxable income until you receive a member equity certificate from the cooperative. At that time you'll make an entry to reverse some of the expense, to convert it to taxable income. See the example later in this section.

- Other deductions, such as for milk hauling, also have negative amounts.

Memorized milk checks?

Because of its complexity, a milk check is an excellent example of the kind of transaction you may want to set up as a memorized transaction. For information about memorized transactions see page 261.

How to Convert Capital Retain Deductions to Taxable Income

A dairy cooperative may make deductions from your milk checks for capital to be retained by the cooperative, an increase in your equity or investment in the cooperative. As you enter milk checks, record any capital retain deductions just like other expense deductions. The deductions accumulate in an expense account (named Capital Retain Deductions in the example earlier in this section).

The IRS considers capital retains to be taxable income to you, but not until you actually receive a certificate of member equity from the cooperative. At that time, transfer the appropriate amount from the capital retain deductions account to the asset account which represents your equity in the cooperative (such as Capital Retain Certificates). Transferring some of the expense account's balance to an asset account accomplishes the job of increasing the farm's taxable income, because afterwards there's less expense to offset income.

You can either use a General Journal entry or the asset account's Register window to make the transfer. The Register window is easier for most people because it uses the words "increase" and "decrease" instead of "debit" and "credit". Here's how to make the transfer entry in the Register window:

1. **Open the Chart of Accounts window.**

   Choosing Lists|Accounts is one way to do this.

2. **Click on the asset account you use to store equity in the dairy cooperative, to highlight it.**

   The account is named Capital Retain Certificates in this example.

3. **With the account highlighted, open the account's Register window.**

   Type Ctrl-R, or click on the Account button at the bottom of the Chart of Accounts window and then click on Use Register in the pop-up menu.

4. **Add a transaction to transfer the appropriate amount from Capital Retain Deductions (expense) to Capital Retain Certificates (asset).**

   Because the Register window represents the Capital Retain Certificates account, enter the dollar amount in the Increase column, and select Capital Retain Deductions in the account field, as the last line of this example shows:

Date	Ref	Payee		Decrease	Increase	Balance
	Type	Account	Memo			
12/31/1998					3,745.00	3,745.00
	GENJRNL	Opening Bal Equity	All prior cap retains			
12/31/1999					833.45	4,578.45
	GENJRNL	Capital Retain Deduction: 1999 Capital retains				

Capital Retain Certificates — How Do I?

About Capital Retains and Revolvements

Capital retains are funds deducted from your milk check and held by a dairy cooperative as part of your investment, or member equity, in the cooperative. They are income you have earned from milk sales, but because they are deducted from your milk check the IRS doesn't consider them taxable income until you receive a certificate of member equity, which shows the dollar amount of capital retained by the cooperative.

Initially you should record capital retains as a deduction from milk sales, just like other expense deductions. When you receive a certificate of member equity from the cooperative, make an entry to transfer an appropriate amount from the capital retains deduction expense account to an asset account representing your equity in the cooperative.

Or at least that's how it ought to work. From the standpoint of the IRS, what's most important is that the amount of taxable income reported on your tax return *matches the amount stated on the 1099 form* you receive from the cooperative. Some cooperatives issue member equity certificates too late to include them in the current year's income reported on the 1099. In these cases, the 1099 form you receive may actually show the amount of capital retained during the cooperative's *prior* fiscal year.

Confusing? Well, don't lose sleep over it. Just remember that the dollar amount you transfer from capital retain deductions (an expense account) to your member equity asset account should match the amount shown on the 1099 form you get from the cooperative, regardless of what's shown on your member equity certificate (except in case of errors, of course).

Depending on the by-laws of the dairy cooperative, some of your member equity may be "revolved" back to you as a cash payment, usually after being held for a number of years. Capital revolvements are not taxable income, because you already paid tax on those funds in prior years—when they became your equity in the cooperative.

How to Enter a Revolvement of Capital Retains

Depending on your cooperative's by-laws, some of your member equity may be "revolved" back to you as a direct payment, usually after being held for a number of years by the cooperative. These capital revolvements are *not* taxable income when you receive them, because you will have already paid tax on them in earlier years (at the time they became equity in the cooperative).

A capital revolvement is just a conversion of one asset type (cooperative equity) to another (money in the farm checking account), which you record when you deposit the revolvement. This example records a capital revolvement by using Capital Retain Certificates as the offsetting account:

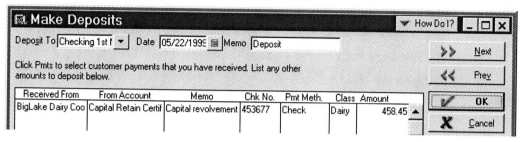

Recording Patronage Dividend Income (Distributions from Cooperatives)

Problem

Usually when I get a patronage dividend from the local cooperative, part of it is distributed to me in cash (a check from the cooperative) and the rest is listed as a "patronage dividend withheld". I know how to enter the cash portion—I just include it in a deposit. But how do I enter the "withheld" portion?

Solution

You can include both the cash and the non-cash portions of a patronage dividend in a deposit entry if you also add a line to offset or "reverse" the effect of the non-cash portion on the checking account's balance.

Discussion

Simply put, the IRS considers the entire amount of a patronage dividend taxable income. You need to enter both the cash and the non-cash portions of a dividend as income so the entire dividend will appear on the income tax reports you print from QuickBooks.

There are several ways you could enter patronage dividend income, but it's best to keep all of the parts of the transaction together as part of the same entry. (That makes it easier to recall the transaction's purpose if you look back at it later.) So a good approach is to enter the non-cash portion of the dividend in the same deposit entry along with the cash portion.

How to Set Up a Patronage Dividend Income Account

If you don't have an income account set up specifically for patronage dividend income, you need to do that before making an entry for the income. Here's how:

1. **Open the Chart of Accounts window.**

   Choosing Lists|Chart of Accounts is one way to do this.

2. **Open the New Account dialog.**

   Either type Ctrl-N, or click on the Account button in the lower part of the Chart of Accounts window and then select New from the pop-up menu.

3. **Set up the account information, as shown here.**

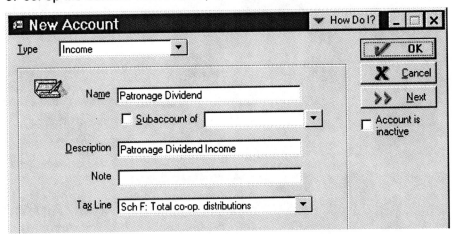

Note that the selection in the Tax Line field chooses where patronage dividend income will appear on QuickBooks' income tax reports. (The choices available in this field may differ depending on the tax form you selected in the "Income Tax Form Used" field of the Company Information dialog, which you can review by choosing File|Company Info.)

How to Set Up an Asset Account for Patronage Dividends Withheld

Though the *non*-cash portion of a patronage dividend is not something you can spend, it is a farm business asset. It is an investment in a business (the cooperative), similar to owning stocks in a corporation. To enter non-cash patronage dividends you must establish an asset account to represent the value of this investment.

1. **Open the Chart of Accounts window.**

Choosing Lists|Chart of Accounts is one way to do this.

2. **Open the New Account dialog.**

Either type Ctrl-N, or click on the Account button in the lower part of the Chart of Accounts window and then select New from the pop-up menu.

3. **Set up the account information, as shown here.**

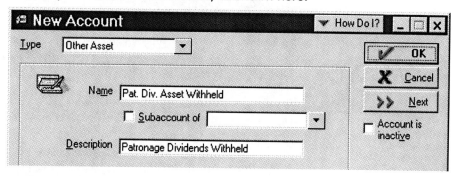

How to Enter Patronage Dividend Income

Here's a deposit entry which records a $44 patronage dividend, half of which has been distributed by the cooperative in the form of a check. The other half has been withheld by the cooperative as a non-cash dividend.

1. **Open the Make Deposits window.**

Choosing Activities|Make Deposits is one way to do this.

2. **Complete the deposit entry as follows.**

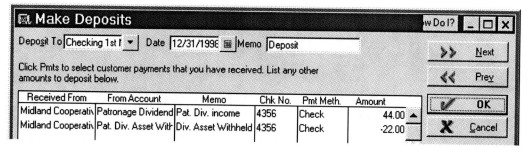

The first line of this transaction records the total amount of the dividend as income. The second line adjusts for the non-cash portion by offsetting $22 of the deposit to the Patronage Dividends Withheld asset account.

What if the Entire Patronage Dividend is Non-Cash?

In some cases, a cooperative may keep all of a declared patronage dividend as retained capital. It wouldn't make sense to record a non-cash dividend in the Make Deposits window, so use the General Journal instead:

1. **Open the General Journal form.**

   Choosing Activities | Make Journal Entry is one way to do this.

2. **Complete the General Journal entry as follows.**

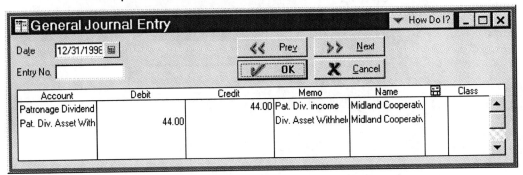

This records both the income and the increase in assets by crediting the Patronage Dividend income account and debiting the Pat. Div. Asset Withheld account.

Tracking Forward Contracts as Pending Cash Sales

Problem

Is there any way for me to keep track of forward contracted sales—both the quantity contracted, and the dollar amount—in QuickBooks?

Solution

You can enter forward contracts as pending cash sales.

Discussion

There are any number of ways to keep track of forward contracted sales—everything from entering them in some other computer program to writing them in a notebook, or on a scrap of paper, or on a wall calendar! But entering contracts in QuickBooks offers several advantages: (1) it gives you one central place to keep the information, where it won't be lost, (2) QuickBooks can remind you prior to each contract's delivery or due date, (3) you can get a report that lists remaining unfilled contracts, and (4) once you've entered the contract in Quick-

Books, only a couple menu clicks are needed to convert it into a transaction record of the actual sale.

The process is simple. Just enter a transaction on the Cash Sales form for the contract quantity and price, and mark it as a Pending cash sale. Later when you actually make delivery on the contract, go back to the cash sales entry and mark it as Final. This converts the original entry, which was inactive (non-posting), into an active cash sale transaction which posts income to your QuickBooks accounts.

Forward contract means a contract for delivery of a specified quantity of grain or some other commodity in a specific future month.

How to Keep Records of Forward Contracted Sales

1. Enter the contract information on the Cash Sales form.

Here's a grain contract example:

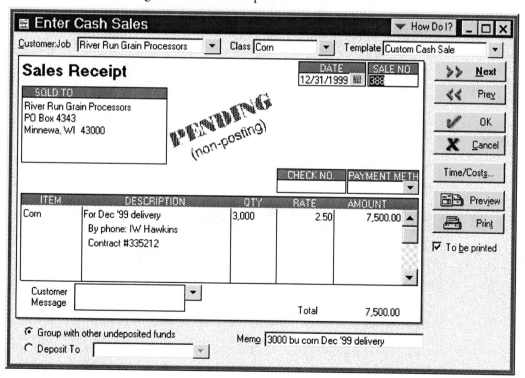

Notes:

* The item you select in the Item column may be any type of item. Typically it will be a Non-inventory Part or Inventory Part type item which

posts revenue to an income account (though no income will be posted while the entry is marked as a Pending cash sale).

- The contract quantity is entered in the Qty (quantity) field

- The Date may be any date you choose. One idea is to always enter the ending date of the contract's delivery period. That way, when you print a Pending Cash Sales report to get a listing of your outstanding (unfilled) contracts, you'll have a handy reminder of when delivery is due on each contract. Another alternative is to always enter the beginning date of the delivery period.

- It's a good idea to type a description of the contract in the Memo field (toward the bottom of the form). This field's contents appears on the Pending Cash Sales report, where it will serve to identify the contract.

2. **With the Cash Sales form still displayed, mark the entry as a Pending cash sale.**

Choose Edit|Mark Cash Sale as Pending. QuickBooks will keep it on file but will not post it to any accounts until you mark it as Final (which removes the Pending designation). QuickBooks displays "PENDING" across the middle of the form, as shown in the illustration above.

3. **(Optional) Set up a To Do note for the contract's beginning and/or ending delivery date.**

QuickBooks can remind you when a contract's beginning and/or ending delivery date is approaching, by displaying a To Do note message in the Reminders window. To enter a To Do note: display the To Do List (choose Lists|To Do Notes), click on the To Do button at the bottom of the list's window, choose New from the pop-up menu, then enter a new To Do note:

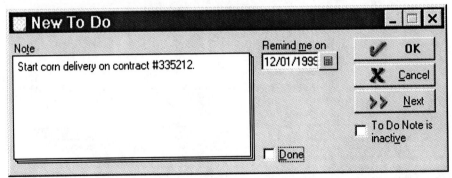

The date you enter in the Remind Me On field is important. That's when QuickBooks will begin listing the To Do note in the Reminders window, assuming you have Reminders turned on. (See page 28 for information on setting Reminders preferences.)

What if a Contract Gets Changed or Canceled?

If the contract quantity or other specifications change, edit the contract's pending cash sales entry. If the contract is canceled, delete the entry.

How to Convert a Contract into an Actual Cash Sale

1. **Find the original contract entry (the pending cash sale).**

2. **Mark the pending cash sale as Final.**

   Choose Edit|Mark Cash Sale as Final. This will allow QuickBooks to post the cash sale normally.

3. **Make other changes, as desired, to properly record details of the sale, then click OK to save the entry.**

   This example records the sale of 3321.45 bushels of corn on the contract entered earlier, including 321.45 bushels that were delivered over and above the contract quantity and sold at a price of $2.43. (Income from the sale will be posted to whichever income account is associated with the Corn item.)

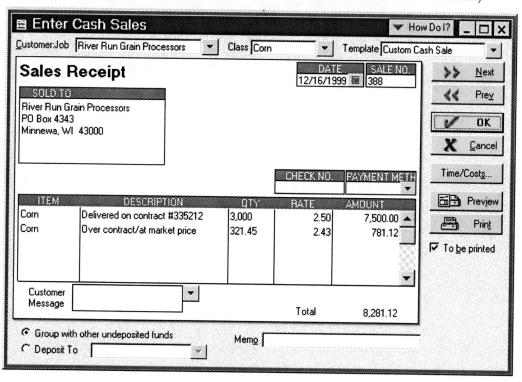

How to Get a List of Outstanding Contracts (Pending Sales)

To get information on outstanding contracts, choose Reports|Sales Reports|
Pending Sales, which will open a report that lists all of the cash sales receipts
marked Pending.

Hint: if you want the contract quantity or other contract information to
appear on the Pending Sales report, enter that information in the Memo
field of pending cash sales entries (the single Memo field at the bottom
of the Cash Sales form, not the Description field).

Your Tax Planning & Tax Preparation Options

Problem

*What's the easiest way to do income tax planning based on my
QuickBooks records? And what about taking records to my tax preparer at tax time?*

Solution

One way is to print an Income Tax report, but QuickBooks
also gives you other options.

Discussion

Informational needs for tax planning and for tax preparation aren't much different. Less accuracy may be required
for tax planning, but in either case you need to make your
records available in a format your tax preparer or a tax software package can use.

Tax Planning/Preparation Options in QuickBooks

With QuickBooks you have three options for tax planning/preparation:

1. **Print an Income Tax report.**

   The Income Tax Summary or Income Tax Detail reports can be printed at
   any time. You may print them before year's end, and take them to your tax
   preparer for filing estimated tax or to decide whether end-of-year purchases
   are warranted to offset some income. You may print them again later, at tax
   time, to use for filing your income tax return. See page 342 for a discussion
   of preparing income tax reports.

2. **Take your records to your accountant, on diskette.**

   The Accountant's Review feature of QuickBooks lets you "check out" a copy
   of your records and take them to your accountant on diskette. The accoun-

tant can open the records on his or her computer, print the necessary report, and even make changes and corrections. When the accountant is done and has returned your corrected records on diskette, QuickBooks will merge the changes back into the master set of records on your computer.

For more information, look up Accountant's Review in the QuickBooks User's Guide and Help system.

3. Buy a tax preparation software package.

You can use tax preparation software both to estimate your tax liability and to prepare your income tax returns. There are a number of tax preparation packages on the market but one of the best at this time is TurboTax, developed by Intuit, the maker of QuickBooks and Quicken.

TurboTax and most of its competitors can *directly* import your QuickBooks or Quicken transactions, saving you some typing and the potential for errors. Based on the tax lines you have assigned to your accounts in QuickBooks, TurboTax will put the imported transaction totals into the proper IRS income and expense categories. After that, you work through a series of questions that guide you through the tax estimation or tax preparation process. When you're all done TurboTax prints a complete set of tax forms on your printer—there's no need to get tax forms from the IRS.

A number of farmers use TurboTax or a similar package with their QuickBooks records each year. Some of these do not actually prepare their own income tax returns. They just use the tax preparation software for tax planning or for filing estimated taxes and let a professional tax preparer file their actual tax return. They justify the cost and time required to learn a tax preparation package by pointing to annual savings of fees their tax preparer would have charged for estimating their taxes. They also note that it's easier to play "what-if" with income taxes when you can plug in different alternatives for asset purchases or prepaid expenses, and see the potential tax effects.

Turbo Tax and similar programs typically cost from $20 to $100 depending on the version you buy—not much money for all they can do—but that's not the end of the story. Because the tax laws change each year you will need to purchase a new copy of these programs annually. Before you buy a tax preparation program be sure it has all the features you need for figuring farm business taxes. Ability to manage a depreciation schedule, for instance, is usually only a feature of tax packages in the upper end of the price range.

See Appendix A for sources and information about tax preparation software products.

Though today's tax preparation software packages are excellent, **always seek the advice of a professional tax preparer when faced with complicated or unusual tax situations!**

CHAPTER 9

Accounting Techniques to "Pick & Choose"

T his chapter describes miscellaneous accounting techniques that are useful in various situations. They won't all apply to *your* situation, but many will, and they are here if you ever need them.

The "Best" Way to Keep Track of Cash Spending & Cash Income

Problem

I'm confused by all the different ways I could handle cash in the farm accounting records. Should I keep a separate cash fund (petty cash account) for the farm? Should all farm cash received be treated as a capital withdrawal for personal use, and "added back" when I use cash to buy something for the farm? What's the best approach?

Solution

This section recommends effective ways to handle cash in your farm business and farm accounting records.

Discussion

Over time we all develop habits for handling farm business cash, and also personal cash. In some cases these habits are mostly accidental; not part of any real plan. That can lead to confusion and poor accounting for cash in the farm's accounting records. Also, if you are new to QuickBooks you may not be aware of certain program features which make it easier to keep good records of cash transactions.

This section provides the information you need for reconsidering your habits for handling cash, in light of what is possible with QuickBooks. Different approaches for handling cash transactions are necessary for each of the three main types of business organization: sole proprietorships, partnerships, and corporations. This section tells you which approaches are "best" for each type.

How to Handle Cash in a Sole Proprietorship

In most sole proprietorships there's a problem with separating "the farm's" cash from "personal" cash. Cash received from selling farm production gets used for personal spending, personal cash gets used to buy repair parts or pay the farm's rural water bill, and so on.

"Who cares?", you might say, "After all, it *is* all mine." And you'd be right. From a practical standpoint, separating farm and personal cash is unimportant. From a management accounting standpoint though, it can be a problem. Mingling farm and personal finances, if done in significant dollar amounts, distorts the picture of farm profitability provided by the farm's accounting records. The remedy is to have a plan for handling cash, and stick to it.

Idea #1: Treat All Cash On Hand as Personal (Non-Farm) Cash

When cash is received in a farm transaction you have two choices. You can either deposit it in the bank or keep the cash to spend. For the cash you keep on hand, usually the best approach is to consider all of it as "personal" cash. This eliminates the need to keep separate "farm" and "personal" cash funds, and therefore prevents confusion over which cash fund should be used.

Here's what you must do to handle all cash as personal cash. When cash is received in a farm transaction, enter it as a personal withdrawal of farm business capital (owner's equity). When cash is paid in a farm transaction, enter it as a personal addition to farm business capital (owner's equity). (Idea #2 tells how to enter these transactions.) Cash is more often spent for personal items. So typically a small number of farm cash purchases will need to be entered.

"Why not issue a farm check to reimburse myself when I make farm purchases with my own cash?"

That's also a good approach but it adds some unnecessary paperwork—extra checks to prepare, to print, and to deposit in a non-farm checking account. For a discussion of reimbursements see page 127, in "Paying Farm Expenses with Non-Farm Funds".

"What if I want to keep a farm petty cash fund?"

QuickBooks makes it easy to keep track of cash spending on behalf of the farm business, using transaction entries alone. Seldom is a farm petty cash fund really necessary. An example of when it may be, is when you need to give employees access to some cash funds for farm business spending. See Maintaining the Farm's Cash Fund (Petty Cash), on page 236, for more details.

Idea #2: Use a Non-Farm Funds account to keep track of cash purchases and withdrawals.

Treating all cash as personal cash (Idea #1) requires entering transactions that not only record cash farm income or expense but also the addition or withdrawal of capital (owner's equity) that results when farm cash is kept for personal use or personal cash is used to make a farm purchase. A Non-Farm Funds account makes this easy.

In QuickBooks, you set up a Non-Farm Funds account using the Bank (checking) account type. That makes it possible to use the account like a checkbook for entering transactions which add or withdraw funds from the farm business.

When cash is received for a farm-related sale and is kept (withdrawn) for personal use, you can enter the transaction as a deposit to Non-Farm Funds. When a farm purchase is made with personal funds, you can enter it as a Check drawn on Non-Farm Funds. (These entries don't correspond to *real* checks or deposits, but using these familiar forms makes the job easy.) Periodically, you record the net amount of capital added or withdrawn by entering a "check" or "deposit" to transfer the Non-Farm Funds balance to an equity account. See page 243 for a full discussion of Non-Farm Funds accounts.

In lieu of using a Non-Farm Funds account, another approach is to directly enter cash farm income and expenses as capital withdrawals/additions, in the General Journal. See pages (for income) and (for expenses) for examples.

Another way to handle cash income is to simply pocket the cash without recording it. It's no secret that income received in small cash payments is grossly under-reported on Federal tax returns. But besides being illegal, omitting large amounts of cash income from the farm business records can affect the profitability picture of your farm business or of a particular enterprise, as viewed by your lender, off-farm partners, or shareholders. It's your decision...

How to Handle Cash in a Partnership

Handling cash in a partnership is basically the same as in a sole proprietorship except that separate accounts are necessary for each partner, to keep track of each partner's additions and withdrawals.

Idea #1: Treat All Partnership Cash as Personal (Non-Farm) Cash

When cash is received in a farm transaction you have two choices. You can either deposit it in the bank or keep the cash to spend. Cash you keep specifically for farm business spending is a petty cash fund. While there's nothing wrong with having a petty cash fund, it may be unnecessary.

The main reason for a petty cash fund is to keep cash on hand to make it available to spend. But typically it's easier for each partner to spend their own cash on farm purchases and simply be reimbursed by the partnership or treat the purchases as additions to farm business capital. QuickBooks makes it easy to keep track of cash spending by each partner on behalf of the farm business, using transaction entries alone. An example of when a petty cash fund may be preferable is when you need to give employees access to cash funds for farm business spending. (See Maintaining the Farm's Cash Fund, on page 236.)

Unless you really need a petty cash fund, a better approach is often to treat all cash as "personal" cash. Cash farm income that isn't deposited in the bank is kept by one of the partners and entered in the farm records as a capital withdrawal by that partner. Each partner also uses his or her own cash to make purchases for the farm. These purchases can be reimbursed from farm checking or entered as capital additions to the farm business. (Idea 2, below, describes the best way to enter these transactions.) Altogether, this eliminates the bother of keeping track of a farm petty cash fund.

This idea is basically the same as described above for sole proprietorships, except that in a partnership each partner's cash purchases and withdrawals must be accounted for separately, using separate accounts.

"What about simply reimbursing partners for the farm purchases they make with their own personal funds?"

That's also a good approach, but it adds some unnecessary paperwork—extra checks to prepare, to print, and to cash or to deposit in the partner's own checking account. For a discussion of reimbursements see page 127, in Paying Farm Expenses with Non-Farm Funds.

Idea #2: Use a Non-Farm Funds Account for Each Partner, to Keep Track of Cash Purchases and Withdrawals.

Treating all cash as personal cash (Idea #1) requires entering transactions that not only record cash farm income or expense but also the addition or withdrawal of capital (owner's equity) that results when a partner keeps farm cash for personal use or makes a farm purchase with his or her personal cash. A Non-Farm Funds account makes this easy.

In QuickBooks, you set up a Non-Farm Funds account using the Bank (checking) account type. That makes it possible to use the account like a checkbook for entering transactions which add or withdraw funds from the farm business. When cash is received in a farm transaction and kept (withdrawn) by a partner, you can enter it on the Make Deposits form. When a farm purchase is made with a partner's personal funds, you can enter it on the Write Checks form. (These entries don't correspond to real checks or deposits, but using these familiar forms makes the job easy.)

In a partnership you should have a Non-Farm Funds account for each partner, to keep track of capital additions and withdrawals separately for each partner. Periodically, you record the net amount of capital added or withdrawn by each partner, by entering "checks" or "deposits" to transfer the Non-Farm Funds account balances to each partner's equity account.

See page 243 for a full discussion of setting up and using a Non-Farm Funds account.

In lieu of using a Non-Farm Funds account, another approach is to directly enter cash farm income and expenses as capital withdrawals/additions, in the General Journal. See pages (for income) and (for expenses) for examples.

How to Handle Cash in a Corporation

The legal requirement for keeping corporate and personal finances separate leads to a different approach for handling cash in a corporation than in sole proprietorships and partnerships.

Idea #1: Maintain a Petty Cash Fund.

A petty cash fund is cash kept on hand specifically for business-related spending. It is an alternative to asking stockholders or employees to make farm purchases with their own personal cash, because the petty cash fund is replenished solely from other corporate funds such as the farm checking account. Having a petty cash fund especially makes sense if you need to give employees access to cash for farm business spending.

Shareholders *should not* use petty cash for personal spending, because that would amount to a personal withdrawal of corporate funds. Shareholders may only withdraw equity from a corporation by selling stock, and should not mingle their personal finances with the corporation's finances. If a shareholder uses cash from the farm's cash fund, he or she should directly reimburse the corporation.

See Maintaining the Farm's Cash Fund (Petty Cash), on page 236, for more information.

Idea #2: Reimburse Stockholders (and Others) Who Use Personal Funds to Make Purchases for the Corporation.

When a stockholder uses personal funds to buy something for the corporation the only "safe" way to handle the transaction is by directly reimbursing the stockholder, usually by issuing a check drawn on the corporate bank account. *Never* treat a farm purchase made with personal funds as a capital addition to the corporation! The only legal way anyone can add to their ownership interest (equity) in a corporation is to purchase stock.

You don't have to issue an individual reimbursement check for each purchase. A better approach is to enter the purchases as bills (Activities|Enter Bills), then periodically use the Pay Bills function (Activities|Pay Bills) to prepare a single reimbursement check for all of them.

Be sure to issue reimbursements on a reasonably frequent schedule, such as monthly. Otherwise it might be possible to make the case that the purchases were a loan to the corporation or constituted a mingling of personal and corporate finances.

Always leave a good "paper trail" for reimbursements.

Both the farm corporation and the person who made the purchase should keep a bill listing the purchased items. Be sure you keep enough detail to show an IRS auditor or a court of law that a check issued as a reimbursement was not simply a way to withdraw corporate funds for personal use.

Maintaining the Farm's Cash Fund (Petty Cash)

Problem

Our farm business keeps its own fund of cash on hand for small purchases and miscellaneous farm spending. Also, when we receive small cash payments, we often add them to the fund rather than depositing them. What's the best way to keep track of these cash income and expense transactions?

Solution

Set up a petty cash account and enter all of the cash fund's transactions in it.

Discussion

The main reason we use cash is convenience. It's often quicker to make small purchases—postage stamps, meals, or a few repair parts—with cash than to write a check. But to have complete tax records and management information you need a record of those cash purchases too, and also of cash received. A petty cash account gives you an easy way to do this.

Petty cash is a small amount of cash kept on hand for miscellaneous spending. It's also an asset, so it should be included on the balance sheet. If your farm business keeps its own fund of cash, then setting up a petty cash asset account in QuickBooks is the best way to keep track of the farm's cash transactions.

QuickBooks makes this job especially easy if you assign the "Bank" account type when you set up the account. That will allow using the familiar Checks and Deposits forms to enter cash income and cash expenses. (These cash transactions aren't really checks or deposits of course...but QuickBooks won't know that!)

This rest of this section tells how to work with farm business petty cash. Farm expenses paid with personal cash or other non-farm funds should be handled differently—see page 127 for details.

Setting Up a Petty Cash Fund and Petty Cash Account

1. **Add a petty cash account in QuickBooks if you don't already have one.**

   Open the Chart of Accounts window (Lists|Chart of Accounts). Then choose Edit|New account to open the New Account window (shown below). Choose "Bank" as the account type. Give the account a name such as Petty Cash, and a beginning balance of zero:

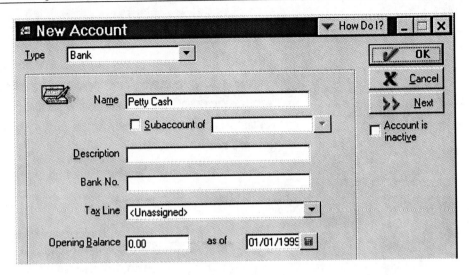

The new account will be grouped with other Bank-type accounts in the Chart of Accounts window.

2. Add cash to the fund so you'll have some cash available to spend.

Before you can spend cash from the petty cash fund, you must of course have some cash on hand! Here are several ways to add cash to the farm business' petty cash fund and properly record the addition in the farm records.

- Withhold some cash from a deposit to the farm checking account. Use Petty Cash as the offsetting account. (See page 152 for an example of withholding cash from a deposit.)

- Supply some of your personal cash to the fund. Enter the transaction by making a "deposit" to the Petty Cash account, using an equity account as the offsetting account (to show that owner's equity was increased in the farm business).

- Receive cash farm income, and "deposit" it to the Petty Cash account (page)

- Write a check on the farm checking account, and cash it. When you enter the transaction use Petty Cash as the offsetting account. Here's an example:

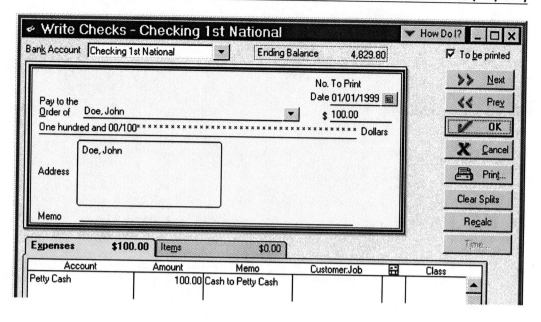

Paying Farm Expenses from Petty Cash

1. **When a purchase is made with farm cash, keep the vendor's receipt or ticket so you'll have a record of the transaction.**

    It's also a good idea to write "CASH" on the receipt so you'll remember it was a cash transaction.

2. **Enter a "check" drawn on the Petty Cash account to record the cash transaction.**

    QuickBooks lets you do this because Petty Cash was set up as a Bank-type account.

    Choose Activities|Write Checks to open the Write Checks window. Be sure to select Petty Cash in the Bank Account field at the top of the window—you don't want to enter the "check" in a real bank account!

    Here's an example "check" drawn on the Petty Cash account:

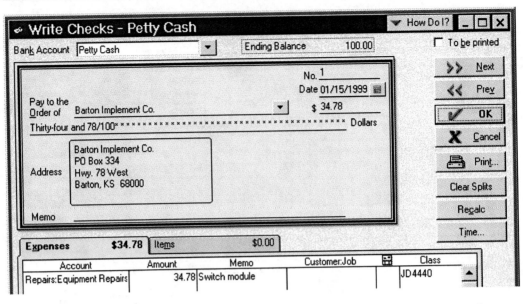

Understand that this entry does not correspond to a real check—there's no check to cash or deposit. The Write Checks form just gives you an easy way to enter cash purchases and assign the proper expense (or other) account.

Receiving Cash Payments to Petty Cash

When you receive a farm-related payment in cash, you may add it to the petty cash fund.

1. Enter the transaction as a "deposit" to the Petty Cash account.

QuickBooks lets you do this because Petty Cash was set up as a Bank-type account.

Choose Activities | Make Deposits to open the Make Deposits window. Be sure to select Petty Cash in the "Deposit To" field at the top of the window—you don't want to enter this deposit in a real checking account.

Here's an example "deposit" to Petty Cash, of cash received from selling hay:

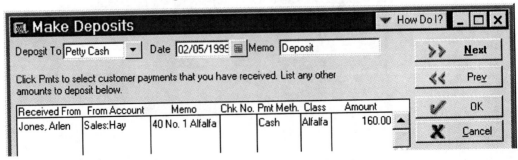

Understand that this entry does not correspond to a real deposit—you won't send it to the bank. The Make Deposits form just gives you an easy way to enter cash payments you've received and assign the proper income (or other) account.

Note that this transaction could just as easily have been entered on the Cash Sales form, if you prefer using items to enter sales. At the bottom of the form, choose the "Deposit To" option to have the cash sale automatically deposited to the Petty Cash account, like this:

Replenishing the Petty Cash Fund from Farm Checking

When you want to increase the amount of petty cash on hand, you may replenish the petty cash fund by cashing a check drawn on the farm checking account, or withholding some cash from a deposit to farm checking. There would be several ways to enter the transaction in QuickBooks: as a check entry for the farm checking account (like the example on page 239), as a "deposit" to the Petty Cash account (with farm checking as the offsetting account), or with a transfer entry (Activities|Transfer Money).

Withdrawing from Petty Cash for Personal Spending

The farm's petty cash fund is a farm asset, and removing any asset for personal use reduces the supply of capital (owner's equity) in the farm business. Your entry needs to show this as an owner's withdrawal of capital.

1. **Enter a "check" drawn on the Petty Cash account, with an equity account as the offsetting account.**

    Here's an example:

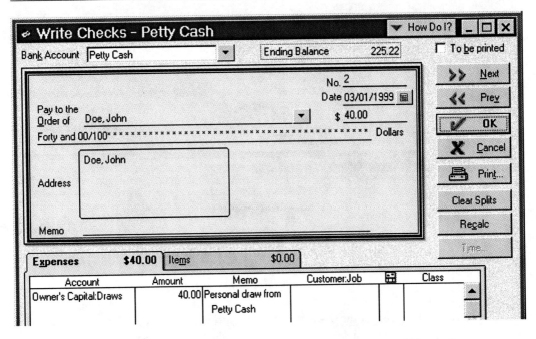

You can select any equity account you want, but it's a good idea to have an account named Capital, etc., available specifically for recording owner's additions and withdrawals of farm capital. (If you don't have a Capital account, add one.)

Partnerships should have a separate capital account for each partner, and assign the appropriate partner's account when he or she withdraws cash for personal use.

Corporations shouldn't allow shareholders to withdraw from petty cash. Shareholders can only withdraw equity by selling stock, and shouldn't mingle their personal finances with the corporation's finances. If a shareholder uses cash from the farm's cash fund, he or she should directly reimburse the corporation.

Reconciling the Petty Cash Account

Reconciling the Petty Cash account from time to time is a good idea, for two reasons:

- You need to periodically verify that the actual amount of petty cash on hand matches the Petty Cash balance in the farm records.

- Reconciling is recommended QuickBooks "housekeeping". If you perform a QuickBooks Condense operation to purge old transactions from your records, transactions in Bank-type accounts will *only* be deleted if they have been reconciled. (Petty Cash, remember, was created as a Bank-type account.)

For details about reconciling see the discussion of reconciling checking accounts, on page 97. Reconciling Petty Cash is mostly like reconciling a checking account, except there's no bank statement to reconcile against.

If reconciling finds a small difference between farm business cash on hand and the Petty Cash balance, and you can't determine the reason, you may click on the Done button in the Reconcile window and let QuickBooks enter an adjusting transaction to correct the account balance.

How often should you reconcile? That probably depends on the number of Petty Cash transactions you enter. Once a year may be often enough in some farm businesses, but reconciling more often can't hurt.

Using a Non-Farm Funds Account

Problem

We frequently make purchases for the farm business from our own funds—using our personal cash or credit card, or a check drawn on our personal bank account. About as often, farm checks and other farm funds are used for personal spending. How do I enter these transactions in the farm records, in a way that properly maintains the farm balance sheet?

Solution

Set up a special-purpose account in QuickBooks, called a Non-Farm Funds account, and assign it the Bank account type. Use it like a checkbook to accumulate transactions that add or withdraw capital from the farm business, such as when you make a farm purchase with personal funds (a capital addition) or use farm funds for personal spending (a capital withdrawal). Periodically transfer the Non-Farm Funds balance to an equity account to record the net change in farm capital (owner's equity) that has resulted from transactions involving non-farm funds.

Discussion

Additions and withdrawals of farm capital most often happen as a side effect of either (1) paying for a farm purchase or farm-related expense with personal funds, or (2) taking cash received as farm income for personal spending. A Non-Farm Funds account makes it easy to track the flow of capital into and out of the farm business, by giving you a place to enter the transactions that involve non-farm funds.

The benefit of setting up a Non-Farm Funds account as a Bank type account is that it lets you enter the transactions involving non-farm funds on QuickBooks' checking account forms, just as if the Non-Farm Funds account were a checking account. You enter farm income on the Deposits form and farm expenses on the

Checks form. Your entries won't correspond to *real* checks or deposits, but using these familiar QuickBooks forms will make it easy to get farm income and expenses posted properly.

Besides entering "checks" and "deposits" in Non-Farm Funds, sometimes you'll use the Non-Farm Funds account as the offsetting account in other transactions. For example, when entering a farm check that will be cashed for personal spending, you would select Non-Farm Funds as the offsetting account for the transaction. Knowing when to use Non-Farm Funds as an offsetting account is easy, because whenever funds "cross over" from farm to personal use or vice versa, Non-Farm Funds is the account you should use.

From time to time (typically before preparing a balance sheet) you will need to transfer the Non-Farm Funds balance to an equity account. But this job is easy too. If Non-Farm Funds has a positive balance you enter a "check" equal to that amount, to zero out the balance. If Non-Farm Funds has a negative balance you enter a "deposit" instead.

Can partnerships use Non-Farm Funds accounts? Sure. Just set up a Non-Farm Funds account for each partner so you can keep a separate record of each partner's withdrawals, and farm purchases made with personal funds.

What about corporations? The owners (shareholders) of a corporation can legally add or withdraw capital only by buying or selling stock. So the only safe way to use a Non-Farm Funds account in a farm corporation is as a ledger of purchases made with non-farm funds and the *direct reimbursements* for those purchases. Consult your accountant before using this approach.

Couldn't I just set up a QuickBooks account for my personal check book, instead of using a Non-Farm Funds account?

You could, if you don't mind having your personal checking account included on the farm balance sheet. But a checking account is not as versatile as a Non-Farm Funds account, because you can't use it to enter transactions involving other types of non-farm funds, like personal cash or credit cards.

How to Set Up a Non-Farm Funds Account

1. Open the Chart of Accounts window.

   Choosing Lists|Chart of Accounts is one way to do this.

2. Add a new account named Non-Farm Funds (or other name of your choice).

Choose Edit|New account to open the New Account window. Choose "Bank" as the account type (that will allow using the Non-Farm Funds account like a checkbook), and assign an Opening Balance of zero. .Here's how the window might look when you're done:

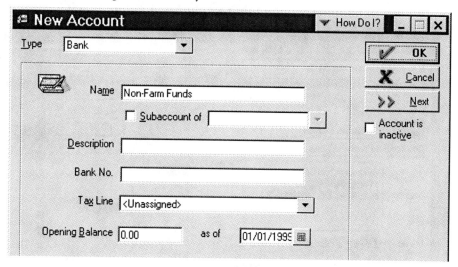

3. Click OK to save the new Non-Farm Funds account.

Advanced technique: multiple Non-Farm Funds subaccounts

It is possible to keep track of different categories of non-farm spending and income by having multiple subaccounts of Non-Farm Funds, one for each category, and using the subaccounts in transactions that involve non-farm spending or income. For ideas on how to set up and work with these subaccounts read about the similar idea of using multiple equity subaccounts, in Keeping Non-Farm Income & Expense Records in the Farm Accounts, on page .

How to Enter a Farm Purchase Made with Personal (Non-Farm) Funds

When you buy something for the farm with funds that are *not* farm business assets—like a personal check or credit card, or your personal cash—you're making a farm purchase with non-farm funds, which you can enter as a "check" in the Non-Farm Funds account.

1. Display the Write Checks form by choosing Activities|Write Checks.

2. Select the Non-Farm Funds account in the Bank Account field at the top of the form.

This step assures that you are entering the check in the Non-Farm Funds account rather than the farm checking account.

3. **Enter the farm purchase or expense.**

Here's an example for a tool purchased with non-farm funds.

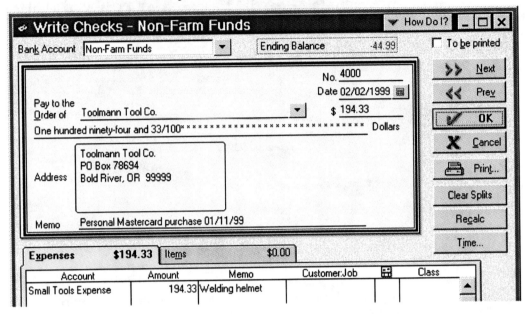

This "check" entry does not correspond to a real check. No check was actually written to Toolmann Tools; in fact, the Memo field tells us the purchase was made by personal credit card. The Write Checks form just provides an easy way to enter purchases made with non-farm funds.

How to Enter Cash Farm Sales when the Cash is Withdrawn for Personal Spending

There are several different things you can do with cash received in a farm sales transaction. You can deposit the cash in the farm checking account (page). You can put the cash in a farm petty cash fund (page) to use for farm-related spending. Or you can keep the cash for personal spending, as a withdrawal from the farm business. If you do the latter, enter the transaction as a "deposit" in the Non-Farm Funds account.

1. **Display the Make Deposits form by choosing Activities|Make Deposits.**

2. **Select the Non-Farm Funds account in the Deposit To field at the top of the form.**

246

This step assures that you are entering the deposit in the Non-Farm Funds account rather than the farm checking account.

3. **Enter the farm income transaction.**

Here's an example of hay sold for cash:

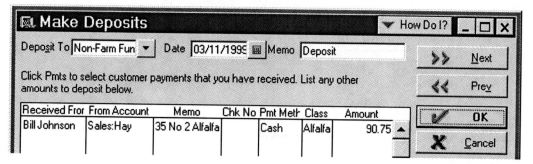

This "deposit" entry does not correspond to an actual bank deposit, of course. The Make Deposits form just provides an easy way to record the income in the Non-Farm Funds account.

When should I enter a "check" in Non-Farm Funds? When should I enter a "deposit"?

Think about the income or expense side of the transaction. If you're entering farm income or a payment received, enter a deposit. If you're entering a farm expense or payment, enter a check.

How to Enter Personal Purchases Made with a Farm Check or Other Farm Funds

When personal items are purchased with farm funds—a farm check, farm petty cash, the farm business' credit card, etc.—use Non-Farm Funds as the offsetting account when you enter the transaction. Here's an example, a farm business check written to pay for household heating fuel:

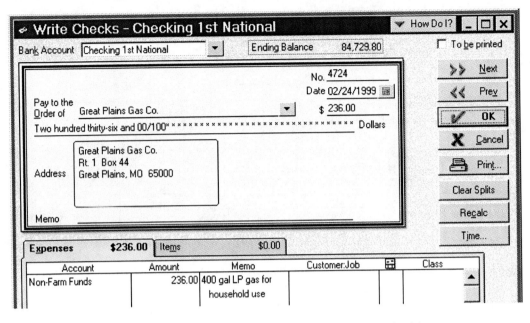

Note that this transaction is entered in the *farm business* checking account, not Non-Farm Funds, because the check was actually drawn on the farm checking account.

 You may use the Non-Farm Funds account when entering *any* transaction that results in an addition or withdrawal of farm business funds.

How to Enter Non-Farm Income Deposited in the Farm Checking Account

When non-farm income is deposited in a farm bank account you're adding funds (capital) to the farm business, so it's appropriate to use Non-Farm Funds as the offsetting account when you enter the deposit. Here's an example, a deposit of Mary Doe's income from substitute teaching at the local high school:

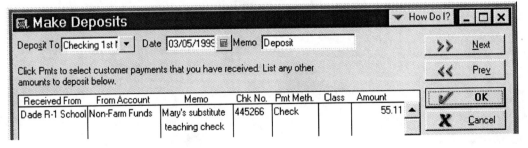

Like the check on the prior page, this deposit is also entered in the farm checking account, with Non-Farm Funds as the offsetting account.

How to Use a Non-Farm Funds Account to Track Purchases and Reimbursements

A Non-Farm Funds account can also be used just for keeping track of purchases made with non-farm funds, for the purpose of issuing reimbursements to the purchasers.

Setting up a separate Non-Farm Funds account for each individual purchaser works well. Purchases made with non-farm funds are entered in those accounts, then periodically checks are issued to reimburse the purchasers. Using the appropriate Non-Farm Funds account as the offsetting account for the reimbursement check entry zeros out the accumulated balance of purchases in the Non-Farm Funds account

 Another way to handle purchases which require reimbursement is to enter them as Bills.

How to "Zero Out" the Non-Farm Funds Balance (Record the Net Change in Equity)

Non-Farm Funds is really a sort of temporary holding tank account that you use for accumulating changes in farm business equity. Over a period of time your "checks" and "deposits" in Non-Farm Funds will leave it with either a positive or a negative balance, which reflects the net change in farm business equity that has occurred. *Before preparing a balance sheet* you need to transfer this Non-Farm Funds balance to an equity account to properly record the equity change, and to zero out the Non-Farm Funds balance so it won't show up on the balance sheet.

 If you only use QuickBooks to keep income and expense records and never use it to prepare a balance sheet, adjusting the Non-Farm Funds account balance is not necessary.

If the Non-Farm Funds balance is negative, it means there is a "shortage" of funds in the account. (The amount of funds withdrawn from the farm business was greater than the amount of non-farm funds added to the farm business.) You need to enter a "deposit" in Non-Farm Funds to cancel out the negative balance, and use an equity account such as Capital or Owner's Equity as the offsetting account. Here's an example:

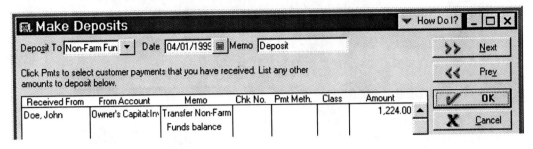

Besides zeroing out the Non-Farm Funds balance this transaction *increases* the balance in the equity account. This reflects a net increase or addition to owner's equity that has resulted from using non-farm funds to make purchases for the farm business.

If the Non-Farm Funds balance is positive, it means there is a "surplus" of funds in the account. (More funds were added to the farm business than were withdrawn.) To cancel out the balance you need to enter a "check" in Non-Farm Funds for an amount equal to the balance, and use an equity account such as Capital or Owner's Equity as the offsetting account. Here's an example:

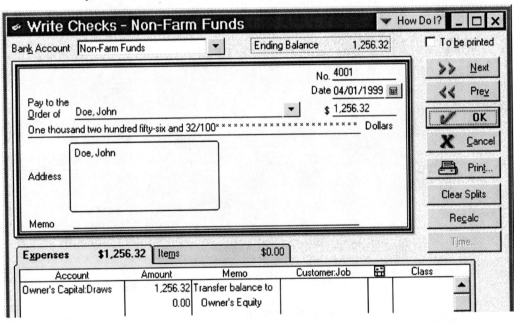

Besides zeroing out the Non-Farm Funds balance this transaction *decreases* the owner's equity balance. A net decrease or removal of owner's equity has resulted from using farm funds for non-farm purchases.

250

If you only use QuickBooks to keep income and expense records and never use it to prepare a balance sheet, transferring the Non-Farm Funds balance to an equity account is not necessary.

Transferring Non-Farm Funds balances in a partnership

A partnership should have a separate Non-Farm Funds account for each partner, and likewise, a separate Capital or Owner's Equity account for each partner. Enter a transaction in each of the Non-Farm Funds accounts to transfer the account balance to the appropriate partner's equity account.

Keeping Non-Farm Income & Expense Records in the Farm Accounts

Problem

We use just one checking account for both farm and personal spending and income. Is there a way to keep personal income and expense records in the same QuickBooks company along with our farm records? If we do that will it be possible to exclude the personal transactions from farm business reports?

Solution

Set up an equity account for each category of non-farm spending or income you want to track. Then when you enter checks for non-farm spending or you deposit non-farm income, assign the spending or income to one of the equity accounts. This keeps the non-farm transactions out of the farm income and expense accounts, yet allows getting a report of personal spending and income.

Discussion

This approach requires thinking of all funds in your Quick-Books accounts as being farm funds—assets of the farm business. That way, spending from the farm checkbook for non-farm items can be treated as a capital withdrawal from the farm business, and non-farm income deposited in the farm account can be treated as a capital addition. Capital additions and withdrawals are entered as increases or decreases to equity accounts. Setting up multiple equity accounts lets you categorize the capital additions and withdrawals, and get separate totals for each category of non-farm spending and income.

The advantage of using equity accounts instead of income and expense accounts, is that it lets you keep track of non-farm spending and income in the farm business records, without actually mingling farm and personal finances! The farm

251

balance sheet will be maintained properly, because non-farm spending and income are recorded as changes in the amount of capital (owner's equity) invested in the farm business.

An alternative to using equity accounts as described here, is to use a Non-Farm Funds account(s). See "Using a Non-Farm Funds Account", on page 243.

Sole proprietorships and **partnerships** may use equity accounts to keep track of non-farm spending and income in the farm business records, but **corporations** should not. The shareholders of a corporation can legally add or withdraw capital only by buying or selling stock.

How to Set Up Equity Accounts for Non-Farm Income & Expenses

1. **Open the Chart of Accounts window.**

   Choosing Lists|Chart of Accounts is one way to do this.

2. **Add new accounts of the Equity account type for the categories of non-farm income and expense you plan to keep track of.**

   Usually it's best to set up these equity accounts as subaccounts of a parent, or controlling, equity account. Your transactions will normally reference the subaccounts, but the parent account will always show the accumulated net total amount of capital added or withdrawn. (QuickBooks automatically includes subaccount balances in a parent account's balance.)

   You can use an existing Equity account as the parent account, or add a new one. To add a new account, choose Edit|New account to open the New Account window. Then choose "Equity" as the account type, name the account as you wish, and give it a beginning balance of zero. Here's an example:

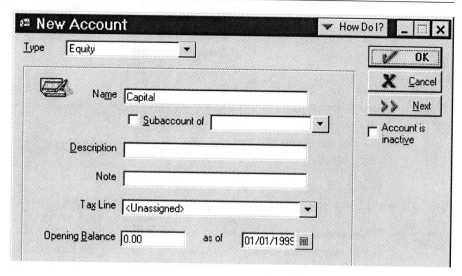

With the parent account in place you can add the detail subaccounts, one for each category of non-farm income and expense you plan to keep track of. Just be sure to check the box beside "Subaccount of" and select the parent account in the space provided. Here's an example:

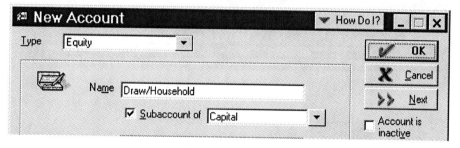

You can of course add as many of these equity accounts as you want, for any categories you desire. But here is an example set to give you some ideas:

> Capital
>> Add/Carpentry Wages
>> Add/Misc.
>> Add/Teaching Salary
>> Draw/Auto Expenses
>> Draw/Electricity (household)
>> Draw/Heating Fuel
>> Draw/Household
>> Draw/Medical Expense
>> Draw/Medical Insurance
>> Draw/Misc.
>> Draw/Other Insurance

Draw/Telephone (non-farm)
Draw/Taxes
Draw/Water (household)

 You can set up the same kinds of equity accounts for a partnership. The only difference is that each partner should have a separate parent account, such as Capital, John Doe, and Capital, Bill Smith. That will let you keep track of each partner's non-farm additions and withdrawals.

Depositing Non-farm Income in the Farm Checking Account

When non-farm income is added to farm business funds, assign the appropriate equity account for capital additions when you enter the income. Here is how wages from an off-farm job might be entered as a farm checking account deposit:

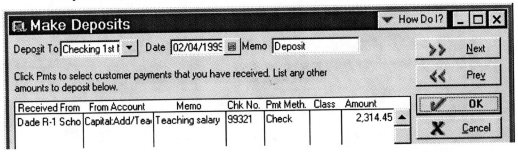

 The examples in this section focus on transactions involving the farm checking account. But you can track non-farm spending and income involving *any* kind of farm funds—checks, cash, farm business credit cards, etc.—by using detailed equity accounts to record capital additions and withdrawals.

Paying Non-farm Expenses from the Farm Checking Account

When a non-farm expense is paid with farm funds, assign the expense to the appropriate capital draw subaccount when you enter the expense. In the following example a farm check is entered for household heating fuel:

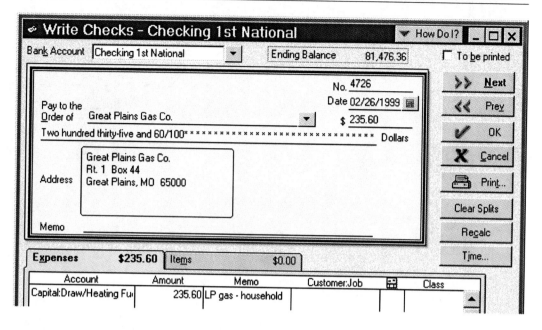

Paying income taxes from farm checking

Knowing how to properly enter income tax payments and refunds is a common problem for new QuickBooks users. Income taxes are a personal expense unless your farm operates as a C corporation (in which case the taxes are a business expense and are paid by the corporation). So in most farm businesses, paying income taxes with a farm check should be treated as a capital withdrawal from the business, by assigning the amount to an equity account, like this:

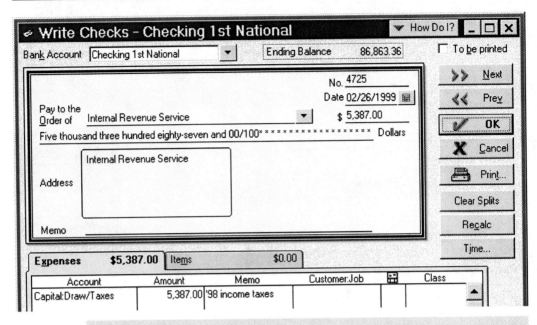

Similarly, **tax refunds** deposited to farm checking are normally capital additions to the farm business. When you deposit a tax refund in the farm checking account assign the amount to an equity account used for recording additions to farm business capital.

How to Get a Report of Non-Farm Spending and Income

The quick way to get a report of non-farm spending and income is to use any of QuickBooks' Balance Sheet reports, filtered to show only the desired accounts.

1. **Open a Balance Sheet report window.**

   Choose Reports|Balance Sheet|Standard, or any of the other Balance Sheet report variations.

2. **Filter the report to include only the desired account(s) and date range.**

   In the report window, click on the Filters button to open the Report Filters dialog. In the Choose Filter area of the Report Filters dialog, select Account as the filter type. Then immediately to the right, select the desired account. In this example, Capital is the parent account under which all non-farm spending and income accounts have been placed, so it is selected as the filter account:

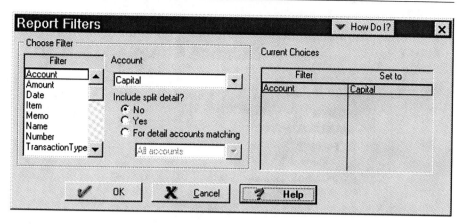

Then set another filter for the transaction dates you want to include (if you don't, the totals may span more than one year's transactions). Click on Date in the Choose Filter area, then immediately to the right of that, select the desired date range.

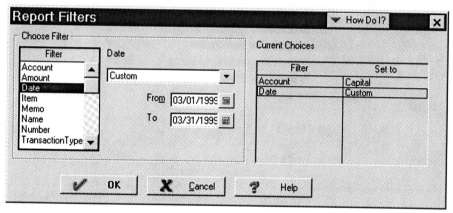

3. Finally, click the OK button to close the Report Filters dialog.

The Balance Sheet report will be redisplayed, limited to Capital and its subaccounts. It should look something like this:

John Doe Farms
Balance Sheet
As of March 31, 1999

04/08/99

	Mar 31, '99
ASSETS	▶ 0.00 ◀
LIABILITIES & EQUITY	
Equity	
Capital	
Add/Carpentry Wages	1,160.47
Add/Misc.	2,241.30
Add/Teaching Salary	2,314.45
Draw/Auto Expenses	-114.23
Draw/Electricity (household)	-48.13
Draw/Heating Fuel	-235.60
Draw/Household	-566.98
Draw/Medical Expense	-24.00
Draw/Medical Insurance	-328.42
Draw/Misc.	-84.12
Draw/Taxes (non-farm)	-5,387.00
Draw/Telephone (non-farm)	-86.88
Draw/Water (household)	-12.50
Total Capital	-1,171.64
Total Equity	-1,171.64
TOTAL LIABILITIES & EQUITY	-1,171.64

How Do I Clear Out the Equity Subaccount Balances?

A minor drawback of using detailed equity accounts for non-farm spending and income records is that their balances are not automatically "closed out" (transferred to another account) at year's end, as income and expense account balances are. Over time, the equity account balances will grow larger and larger if you don't manually make offsetting entries to zero them out.

Probably the best approach, is to transfer the equity subaccount balances to the parent account, with General Journal or Register entries. For example, suppose you have this simple set of equity accounts and balances:

258

Capital	10,520.30
Add/Teaching Salary	18,346.55
Draw/Auto expenses	-3,371.00
Draw/Household	-4,455.25

The Capital:Add/Teaching Salary subaccount is carrying a credit (positive) balance, and the two drawing subaccounts are carrying debit (negative) balances. Here is a General Journal entry to transfer the subaccount balances to Capital, the parent account:

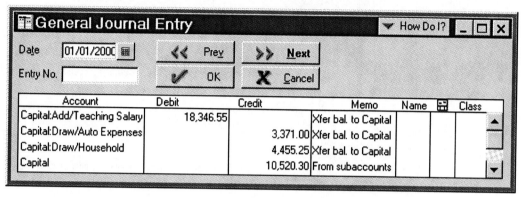

Important! This General Journal entry must be dated properly. Always date the entry using the *first day of your accounting year* (fiscal year). That returns the subaccount balances to zero as of the beginning of each year, while still allowing reports of non-farm spending and income to be obtained for prior years (by getting a balance sheet report filtered for the last day of a specific year).

Here are the resulting account balances following the General Journal entry:

Capital	10,520.30
Add/Teaching Salary	0.00
Draw/Auto expenses	0.00
Draw/Household	0.00

Note that the General Journal entry had no *net* effect on the Capital account's balance! Transferring the subaccount balances merely rearranged the balance sheet—it did not change the net amount of farm business equity.

Automating Recurring Transactions

Problem

I enter some transactions every month, almost exactly as I've entered them in the prior month—paying the monthly electric bill is a good example. Is there any way QuickBooks can make these entries automatically, so I won't have to type so much?

Solution

QuickBooks has two features which can help reduce the typing needed for entering recurring transactions: automatic recall, and memorized transactions.

Discussion

Recurring transactions are ones you enter over and over again—every month, every quarter, or every year. Everyone pays some bills on a recurring schedule, such as the electric bill, water bill, rent or lease payments, and loan payments. Or maybe you send out basically the same bill to all of your customers, over and over again, month after month.

QuickBooks gives you at least two ways to enter recurring transactions without a lot of typing:

- *Automatic recall* is a feature which helps complete a transaction you've begun entering, by reusing details from a recent transaction involving the same person or business (vendor or customer).

- *Memorized transactions* let you record specific transaction details and recall them to enter a particular type of transaction. You can even schedule memorized transactions for automatic entry, which means QuickBooks will create selected transaction entries for you on a repeating schedule which you choose.

By the way, you may use both automatic recall and memorized transactions—you aren't limited to using just one or the other. And both features work on most QuickBooks forms, including checks, bills, cash sales, invoices, and others.

Automatic Recall

Automatic recall is simple and easy because it's truly automatic. When you select a customer or vendor name in a QuickBooks form, QuickBooks automatically fills in the rest of the form with details from the most recent transaction entered for that customer or vendor. This usually saves lots of typing—you just edit the details as necessary for the new transaction, and move on!

Automatic recall works especially well for transactions you enter basically the same way each time. The monthly electric bill would be a good example.

All that's required to use automatic recall is to turn it on in the Preferences window:

1. **Choose File|Preferences to open the Preferences window.**

2. **Click on the General icon in the scrollable box along the left side of the window.**

   You may have to scroll the box up or down to find it.

3. **Select the "Automatically recall last transaction for this name" check box.**

4. **Click OK to close the Preferences window.**

Automatic recall only helps when *you* choose to enter a transaction. It cannot enter transactions automatically. For that, you need to use memorized transactions.

Memorized Transactions

Some users depend heavily on the memorized transactions feature and find it very useful; others feel it gets in their way. But in any case, it deserves a look if you enter some transactions in roughly the same form each time you enter them. Setting up memorized transactions takes only a little time and can pay off later with big time savings.

You can also select some memorized transactions for automatic entry. QuickBooks will automatically enter those transactions at predetermined intervals—once a month, once a year, etc. Besides reducing the amount of typing you must do, scheduled transactions assist in getting bills paid on time: the scheduled bills automatically appear in the Reminders window when it's time to pay them.

How to Create a Memorized Transaction

Think of a memorized transaction as set of instructions which tells QuickBooks how to fill out a form. To create a memorized transaction you first create a model transaction by filling in a form's fields with the information you want to memorize. Then you tell QuickBooks to memorize the transaction—to save a copy of the model in the Memorized Transactions list. Here are the basic steps for memorizing a transaction:

1. **Open the QuickBooks form in which you want to memorize a transaction.**

   Each memorized transaction is particular to a specific QuickBooks form.

2. **Create a model transaction by filling out the form with the details you want QuickBooks to memorize.**

When you recall a memorized transaction, QuickBooks will use *all* of the information you've memorized. So be selective about what you include in the model transaction. Fill in the fields which likely will be the same each time the transaction is entered, such as account and class information, and maybe the customer or vendor name. *Do not* fill in fields which will be different each time, such as the check number, check or bill total, or the dollar amount assigned to each account. (Don't worry about the date. QuickBooks supplies the current date when you recall a memorized transaction.)

Here's a model check for memorizing the monthly payment to the electric company. Note that it pre-selects accounts and classes for splitting the electric bill three ways: between a hog enterprise class, general farm use, and household electricity use:

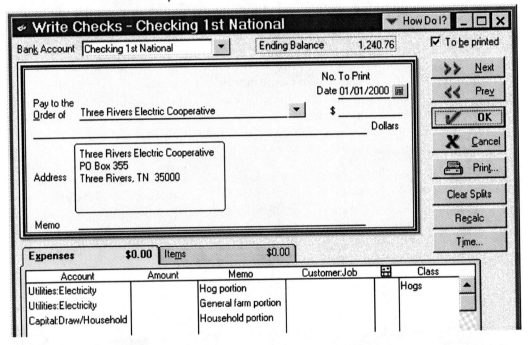

A note to Quicken users...

In Quicken, memorized transactions can allocate the transaction total among several accounts (splits) on a percentage basis. But memorized transactions in QuickBooks only "remember" exact dollar amounts. That's why QuickBooks transactions should usually be memorized without dollar amounts. The exception is transactions which are normally entered for the same amount each time.

3. Memorize the model transaction.

Choose Edit|Memorize *formname*, where *formname* is the name of the form you are memorizing. For example, to memorize a check the menu item would read: Edit|Memorize Check.

QuickBooks will display a Memorize Transaction dialog which lets you choose options for how to memorize the transaction.

4. **Select options as desired for memorizing the transaction.**

Here is the Memorize Transaction dialog, along with descriptions of some of its fields:

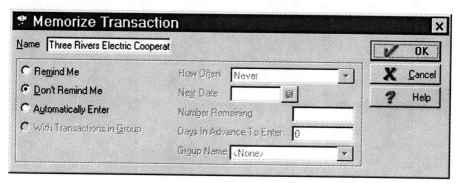

Name

The name you enter here will identify the transaction in the Memorized Transactions list. You may use the name QuickBooks suggests—usually a duplication of the Customer or Vendor field—or change it to any name you want.

Remind Me/Don't Remind Me

These two options control whether QuickBooks will automatically remind you when to use memorized transactions, on a repeating schedule. (See the Reminders window discussion, on page 29.)

Automatically Enter

Choosing this option allows setting up a memorized transaction for automatic entry by QuickBooks, on a schedule you determine (discussed later in this section).

With Transactions in Group

This option is for including the memorized transaction in a group with others, and is only available if you have created one or more transaction groups (discussed later in this section).

Other fields in the Memorized Transaction dialog are for setting up reminders, scheduling, and transaction groups, all of which are discussed later in this section.

5. **Click OK to close the Memorize Transaction dialog.**

QuickBooks will add the model transaction to the Memorized Transactions list.

6. **Cancel the model transaction.**

Unless the model transaction is one you actually want to save, click on the Cancel button in the form's window to prevent saving it.

You can memorize existing transactions, too...

Find the existing transaction you want to use as a model and remove any information you don't want to memorize, then memorize the transaction. After memorizing it, be careful not to save the changes you 've made in the transaction (.e., in case you removed important information to make the transaction into a model). Click the form's Cancel button to avoid saving the model transaction in place of the original.

How to Manually Recall (Use) Memorized Transactions

There are actually several ways to recall and use memorized transactions. You can manually select a memorized transaction to use, or have QuickBooks remind you about using the transaction at regular intervals such as once a month, or have QuickBooks automatically enter the transaction on a regular schedule—monthly, quarterly, annually, etc. Reminders and scheduling of memorized transactions are discussed later. For now, here are the manual steps for using a memorized transaction.

1. **Open the Memorized Transactions List window.**

Choose Lists | Memorized Transactions to open the window:

Memorized Transaction List ▼ How Do I?

Transaction Name	Type	Source Account	Amount	Frequency	Auto	Next Date
◈ Land payment - Mrs. Bell	Check	Checking 1st National	0.00	Annually		03/15/2000
◈ Pasture Rent - Bill Johnsc	Check	Checking 1st National	0.00	Twice a year		10/01/1999
◈ Stone Co. Rural Water St	Bill	Accounts Payable	0.00	Monthly		04/01/1999
◈ Three Rivers Electric Coc	Check	Checking 1st National	0.00	Never		

Memorized Transaction ▼   Enter Transaction

2. **Select the memorized transaction to enter.**

Either double-click the desired memorized transaction, or highlight it and then click the Enter Transaction button in the lower part of the window).

QuickBooks will enter the transaction by opening the appropriate form and filling in the fields as they were memorized. (Each memorized transaction is specific to the QuickBooks form where it was created.)

3. **Fill in the rest of the form, correcting and adding information as necessary, then save the new transaction entry.**

Recalling a memorized transaction merely fills in the memorized fields. You may change or correct any fields, as needed, before saving the transaction. Here's a completed check entry created by recalling the Three Rivers Electric Cooperative memorized transaction and adding dollar amounts:

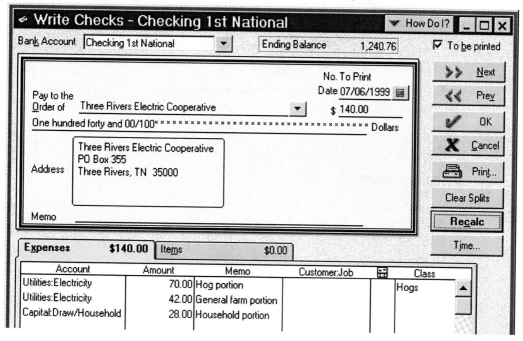

How to Change the Information Stored in a Memorized Transaction

To change the information stored in a memorized transaction, you need to re-memorize the transaction. Here's how:

1. **Recall the memorized transaction (as described on the prior page), as if you were going to enter a new transaction.**

QuickBooks will open the appropriate form and fill in the memorized information.

2. **Make changes in the form's fields, as desired.**

3. Choose Edit|Memorize *formname*, where *formname* is the name of the form you are memorizing.

   This is the same step you used to originally memorize the transaction. But this time QuickBooks may recognize that a transaction with the same name is already present in the Memorized Transactions list. If so, it asks if you want to replace the old memorized transaction or add a new one:

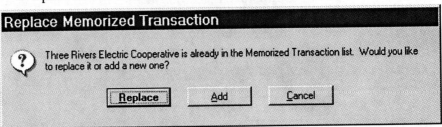

4. Click the Replace button to save the changed transaction information in place of the old.

Having QuickBooks Remind You When to Use Memorized Transactions

You can have QuickBooks regularly remind you when it's time to use a memorized transaction. For example, you can get a reminder about paying the rural water bill at the beginning of each month. Here's how to set up a memorized transaction so you will be reminded about it:

♦ When you memorize the transaction, select the Remind Me option and enter the desired reminder settings.

   Hare are some of the field choices you may enter:

   How Often
   Select an option such as Monthly, to indicate how often you want to be reminded to use the transaction.

   Next Date
   Enter a date here to indicate when you want to begin being reminded about the transaction.

   Here's an example of the reminder settings for a water bill payment:

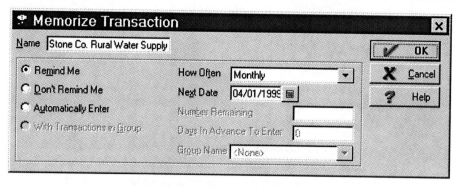

Based on these settings, QuickBooks will display a reminder about using this memorized transaction, beginning on the first day of each month, in the Reminders window. (See page 28 for information about turning on the Reminders feature, and related settings.)

 You can also change these settings for a transaction you've already memorized, by selecting the transaction in the Memorized Transaction List window and choosing Edit|Edit Memorized Transaction.

Scheduling Memorized Transactions for Automatic Entry

You already know that memorized transactions work well for entering bills and other transactions which repeat regularly. Why not take the idea one step farther and let QuickBooks actually enter those repeating transactions for you?

Scheduling automatic transaction entries is a powerful technique...so use it carefully. A good practice is to always check over the auto-entered transactions to be sure they contain the information you intend.

The best candidates for automatic entry are transactions which have a "To be printed" check box, such as checks, bills, and invoices. If your automatic entries include a check mark in the "To be printed" box, the Reminders window can show which ones are not yet printed. In effect, this marks the automatic entries as transactions you need to check for errors, which you can do before printing them.

Here's how to set up the Stone Co. Rural Water Supply District memorized transaction, described on the prior page, for automatic entry on a monthly schedule:

♦ **When you memorize the transaction, select the Automatically Enter option and enter the desired settings.**

Here are some of the field choices:

How Often
Select an option such as Monthly, to indicate how often you want the transaction to be entered.

Next Date
Enter the first date on which you want QuickBooks to begin making the automatic entries.

Number Remaining
If you only want the automatic entries to be made a specific number of times, enter that number here—you might enter 12 for twelve monthly loan payments, for example. Otherwise leave the field blank.

Days in Advance to Enter
If you've entered a payment's actual due date in the Next Date field, you may enter the number of days ahead of that date to generate the automatic entry. Otherwise leave the field blank.

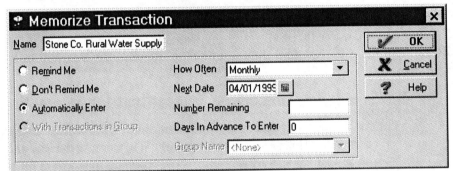

You can also change these settings for a transaction you've already memorized, by selecting the transaction in the Memorized Transaction List window and choosing Edit|Edit Memorized Transaction.

With the memorized transaction set up this way QuickBooks will automatically enter a transaction based on it, on the first day of every month.

Each time you start QuickBooks or open a different company file QuickBooks checks for automatic entries that need to be made. If it finds any, a dialog is displayed asking if you want to go ahead with making them. This alerts you to the fact that new automatic entries have been made, so you can check them to assure they are correct.

A few more comments...

- If you don't happen to start QuickBooks on the date of a scheduled automatic entry, QuickBooks will make the entry the next time you start up the program.

- Always review automatically entered transactions to be sure they contain the information you intend.

- If you decide not to use a transaction that QuickBooks has automatically entered, just delete it. QuickBooks will auto-enter the transaction again next month.

- To stop auto-entry of a transaction, open the Memorized Transactions list and delete it.

A Tricky Job: Specifying "Terms" for Memorized Bills

The Terms field is crucial for determining the due date of bills that are automatically entered by QuickBooks from the Memorized Transactions list.

This one is important for determining the due date of scheduled transactions automatically entered by QuickBooks. Say you define terms as Net 10 (due 10 days from the entry date) in the memorized transaction. Further, suppose you don't use QuickBooks every day—maybe you haven't started QuickBooks for five days or so when the automatic entry should have been entered.

When QuickBooks makes the automatic entry, what will it do? If the automatic entry isn't entered until the 30th day of the month, when it should have been entered around the 20th, the due date for the auto-entered bill will be 10 days later than it should be—because the Terms field dictates a calculated due date 10 days after the entry date.

If you only start up QuickBooks once a week or so, what can you do to prevent paying bills too late? Consider setting up some "Date Driven" terms (in the Terms field) for memorized transactions that will be automatically entered. Here's how:

1. Click the down arrow in the Terms field, and select <Add New> to add a Terms definition.

   QuickBooks displays the New Terms dialog. Here's what it should look like after making the entries described below:

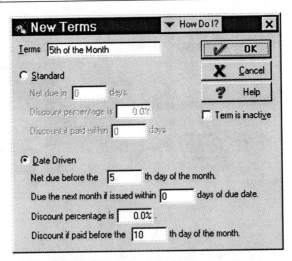

2. Enter a descriptive name for the new definition, in the Terms field.

3. Click on Date Driven to enable the Date Driven fields.

4. Enter a number for the day of the month that will be the due date.

   The 5 used in the New Terms dialog here corresponds with the name "5th of the Month". On any bill using this Terms definition the due date will be the next available date that is the 5th day of the month. Bills entered on the 20th of February, 2nd of March, or 5th of March would all have due dates of March 5th. But a bill entered on March 6th would have a due date of April 5th.

5. Enter a discount percentage and discount day of the month, if they apply.

6. Click OK to save the new Terms definition.

Using this Terms definition in a Memorized Transaction for a monthly bill will cause each automatic entry of the bill to have a due date of the 5th of the month, even if the transaction gets entered a few days later than normal.

Grouping Memorized Transactions

QuickBooks lets you group memorized transactions together and handle them as a unit. This is a powerful feature, because selecting a single group entry in the Memorized Transactions List window can cause QuickBooks to enter any number of transactions—as many as you've included in the group.

The most common reason to group memorized transactions occurs in businesses and organizations that need to regularly bill customers or members for a specific dollar amount, on a recurring basis. For example, many rural fire protection associations annually bill their members for membership dues. Other businesses bill for lease or rental payments, monthly service fees, etc.

In a farm business there are also ways to benefit from grouping some of your memorized transactions. For example, you probably have several bills which are normally paid during the first week or so of each month—electric bills, water bills, and so on. Those bills could be scheduled for automatic entry, as described earlier in this section. But if you are uncomfortable with having bills entered automatically, an alternative is to create memorized transactions for each of the bills and include them in a memorized transaction group—name it something like "First of the Month". Entering the entire batch of bills each month then becomes as simple as selecting the "First of the Month" group in the Memorized Transactions List window, and clicking on the Enter Transaction button.

You can add a memorized transaction to a group when you memorize it, or later by selecting the group entry in the Memorized Transaction List window and editing its information. See the QuickBooks User's Guide or Help system for details.

Managing QuickBooks Lists: Reordering, Merging, Deleting, and Hiding List Entries

Problem   *Can I change the order of my Chart of Accounts list? What about removing old Customer and Vendor names I no longer use?*

Solution   This section details several techniques for managing QuickBooks lists.

Discussion   With few exceptions, the techniques described below apply to all lists: Accounts, Classes, Items, Customers, Vendors, and others.

Changing the Order of List Entries

QuickBooks basically gives you two ways to reorder any list. You can have QuickBooks sort the list alphabetically, or you can manually move list entries where you want them by dragging them with the mouse.

To sort a list, first open the list's window. Then choose Lists | Resort *listname* list, where *listname* is the name of the list. QuickBooks will proceed to rearrange the list in alphabetical order.

To change the order of list entries by dragging them, move the mouse cursor directly over the little diamond shape that appears just to the left of the entry you want to move. When the cursor changes to a cross shape, click and hold down the left mouse button and drag the item upward or downward in the list.

You can also change the "level" of list entries in the list hierarchy. Suppose you want to make one account a subaccount of some other account. Move the mouse cursor over the diamond shape as before, but this time drag to the right to indent the account below another account. You can also drag to the left, to change a subaccount into a higher-level subaccount or a main account.

List entries can move far!

Sometimes if you change the level of a list entry (change its position within the list hierarchy), QuickBooks may move it quite a distance from its original location, upward or downward, due to automatic sorting of list names. If this happens, just drag the member up or down to reposition it where you want it.

What to Do with List Entries You No Longer Use

When any QuickBooks list becomes cluttered with items you no longer use—old accounts, customers to whom you no longer sell, etc.—you may delete them, hide (inactivate) them, or merge them with another list entry.

Deleting List Entries

Deleting entries from a QuickBooks list is rarely an option, because QuickBooks will only let you delete list members that are not used in any way in your transactions. For example, deleting an account is only permitted if (1) it has no subaccounts, and (2) it is not used in any transaction that still exists in your Quick-Books records, and (3) it is not assigned to any item in the Items list.

To delete a member from a list:

1. **Open the list's window.**

Choose Lists|*listname*, where *listname* is the name of the list, to open the window.

2. **Highlight the list item you wish to delete, by clicking on it.**

3. **Delete the list item.**

You can either type Ctrl-D, or choose Edit|Delete *listitem*, where *listitem* is the name of the list type, like Account, Class, Customer, etc.

QuickBooks will display a message if the list entry cannot be deleted.

Hiding List Entries

This is a quick way to get unused list entries out of the way. They're not actually removed from the list, just marked inactive. Then in QuickBooks forms where you can select something from the list, the inactive members aren't shown. Having fewer members visible in a list makes it easier to use.

To hide (inactivate) a list entry:

1. **Open the list's window.**

Choose Lists|*listname*, where *listname* is the name of the list, to open the window.

2. **Highlight the list item you want to hide, by clicking on it.**

3. **Mark the list item inactive.**

One way to do this is by choosing Edit|Mark listitem Inactive, where listitem is the name of a list entry like Account, Class, Customer, etc.

 You can toggle the Show All check box at the bottom of the list window, by clicking on it, to hide or show the inactive list entries.

Merging List Entries

Marking unused list entries as inactive hides them, but the are still be present in the list. Merging list entries is a way to completely remove them from the list, even if they are in use somewhere in your records.

Some inventory techniques require adding new inventory items to the Items list frequently—maybe dozens of times each year. A good way to keep the list size manageable in this situation is to merge old, unused items under one item name. Suppose part of the Items list contains the following item names:

Name	Description	Type	Account	On Hand	Price
◇ 96st1023-1	BWF strs @ 588 lbs	Inventory Part	Sales:Resale Livesto	0	0.00
◇ 96st1023-2	BWF strs @ 553 lbs	Inventory Part	Sales:Resale Livesto	0	0.00
◇ 97st0918-1	Red angus strs @ 5	Inventory Part	Sales:Resale Livesto	0	0.00
◇ 97st0918-2	Blk strs @ 543 lbs...	Inventory Part	Sales:Resale Livesto	0	0.00
◇ 97st1003-1	Blk strs @ 514 lbs	Inventory Part	Sales:Resale Livesto	0	0.00
◇ 97hf1003-2	Blk hrs @ 499 lbs....	Inventory Part	Sales:Resale Livesto	0	0.00
◇ 98st1015-1	Black strs @ 503 l...	Inventory Part	Sales:Raised Livesto	0	0.00
◇ 98st1015-2	BWF strs @ 544 lbs	Inventory Part	Sales:Raised Livesto	0	0.00
◇ 99st1122-1	RWF strs @ 546 lbs	Inventory Part	Sales:Resale Livesto	52	0.00
◇ 99st1122-2	Char X strs @ 592 ..	Inventory Part	Sales:Resale Livesto	41	0.00
◇ ZZResaleLvstk	(Archival) Resale ...	Inventory Part	Sales:Resale Livesto	0	0.00

Item ▼   Activities ▼   Reports ▼   ☐ Show All

These are items you might create for groups of feeder cattle purchased for resale, as discussed on page 199. The first two characters of each item name indicate the year of purchase for the various groups, so it's easy to tell which items are old ones and no longer active. (The remainder of the name indicates the sex of the cattle, the month and day of purchase, and a lot number.)

QuickBooks won't allow deleting old items as long as transactions exist which still use them. But you can merge the old items with other items you've set up specifically for archival purposes—like the ZZResaleLvstk item in the Items list above. (The item was named with "ZZ" at the beginning so QuickBooks will get it out of the way by sorting it toward the bottom of the Items list.)

One way to remove some of the items, like 96st1023-1 and 96st1023-2, from the Items list would be to merge them with ZZResaleLvstk. Yes, some detail would be lost from transactions where 96st1023-1 and 96st1023-2 were used, but the transactions would still contain all of the other original details. And you can retain as much or as little detail in archival items as you want. For instance, setting up separate ZZResaleStrs and ZZResaleHfrs archival items would keep steer and heifer transaction details separate.

The following procedure shows how to merge the 96st1023-1 item into the ZZResaleLvstk item. (Merging can be done in any QuickBooks list, including the Accounts and Classes lists, using the same steps.)

Warning! Merging list entries cannot be undone!

Wait until a list entry is a year or two into history before merging it with another. Until then, just hiding the list entry by marking it inactive would be a safer choice.

1. Open the Items list.

274

Choosing Lists | Items is one way to do this.

Note: list entries must be at the same level in the list's hierarchy before you can merge them. (List entries on the same level are indented the same number of spaces in the list window.) If 96st1023-1 and ZZResaleLvstk were at different levels in the Items list, you would need to drag one of them to the same level as the other before continuing.

2. Highlight the 96st1023-1 item, by clicking on it.

3. Open the 96st1023-1 item's edit window.

Typing Ctrl-E or choosing Edit | Edit Item is one way to do this.

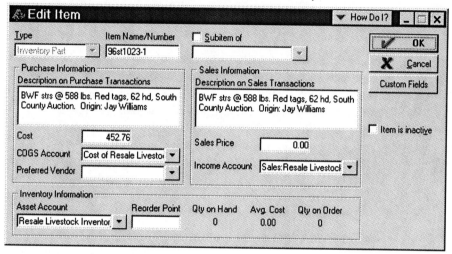

4. Change the item name (in the Item Name/Number field) to "ZZResaleLvstk".

In other words, change the name of the old item to the name of the item you want to merge it with.

5. Click OK.

QuickBooks will display a message asking if you want to merge the items.

6. Click Yes to confirm that you want to merge the two items.

Converting from Enterprise Subaccounts to Enterprise Classes

As described elsewhere in this book, using QuickBooks classes is the best way to identify enterprise information in your transactions. But what if you've set up your Chart of Accounts the other way—using accounts and subaccounts to identify enterprises. For example, suppose your chart of accounts contains this Fertilizer account, with subaccounts for identifying fertilizer costs for different crops:

Fertilizer
Corn
Soybeans

If you switch to using classes you can simplify the chart of accounts to have just one Fertilizer account, with no subaccounts:

Fertilizer

And the Class list would identify the crop enterprises:

Corn
Soybeans

But with a lot of transactions already entered based on the "old" chart of accounts setup, how do you switch to using classes? The job may require a considerable amount of work if you have many transactions, but it is possible to make the switch to classes without losing any of the enterprise information you've already entered. Here's how.

First, add the classes to the Class list.

1. **Open the Class list window.**

   Choosing Lists | Classes is one way to do this.

2. **Open the New Class dialog.**

   Either type Ctrl-N, or click on the Class button in the lower part of the Class list window and then select New from the pop-up menu.

3. **Set up the new classes.**

Second, add class information to the transactions that currently use the Fertilizer:Corn and Fertilizer:Soybeans subaccounts.

1. **Open the Chart of Accounts window.**

   Choosing Lists | Chart of Accounts is one way to do this.

2. **Click once on the subaccount you want to remove, to highlight it.**

3. **Click on the Reports button at the bottom of the Chart of Accounts window.**

4. **From the pop-up menu choose QuickReport.**

   QuickBooks will open a report window containing transactions for the highlighted account.

5. **Filter the report to include all dates.**

   Click the down arrow in the Dates box near the top of the report window, and select All (all dates).

6. **Customize the report to include the Class column (if that column is not already included in the report).**

   Click on the Customize button at the top of the QuickReport window to display the Customize Report dialog. In the Columns box along the left side of the Customize Report dialog, scroll down until "Class" is visible, then click on it to place a check mark beside it. Click OK to close the Customize Report dialog.

    If no Class column is listed in the Columns box of the Customize Report dialog, you need to turn on the Class tracking preference as described on page .

7. **Use the QuickReport window to locate each transaction for editing, one by one, and add class information to them.**

   When you move the mouse cursor over a transaction listed in the report, the cursor will change to a magnifying glass shape. When it does, double-click the left mouse button to open the transaction in the appropriate form. Modify the transaction by selecting the desired class in the Class field, then close the form by clicking its OK button.

   Repeat this action for all transactions listed in the QuickReport window...yes, this may take quite a while if you have many transactions! When you are done adding class information to the transactions, close the QuickReport window.

8. **Repeat steps 2 through 7 for each subaccount you want to convert to classes.**

Third, mark the subaccount(s) as inactive.

The purpose of this step is to hide the subaccounts so you won't accidentally use them in the future when entering transactions.

1. **Open the Chart of Accounts window.**

   Choosing Lists|Chart of Accounts is one way to do this.

2. **Click once on the subaccount you want to mark inactive, to highlight it.**

3. **Click on the Account button at the bottom of the Chart of Accounts window.**

4. **In the pop-up menu, click on Make Inactive.**

In the Chart of Accounts list a special symbol appears to the left of the accounts you have marked as inactive, to let you see which accounts are inactive. Those accounts will remain in the Chart of Accounts list but will not be available for selection when you enter new transactions, unless you remove the inactive marking—which you may also do from the Chart of Accounts window.

Another way: merging accounts

An alternative to hiding inactive subaccounts is to merge them with their parent accounts. This causes QuickBooks to actually change all the transactions where the subaccount was used. For details on merging list entries (which includes accounts) see the discussion on page 273, and also the QuickBooks Help system.

CHAPTER 10

Farm Asset Basics

Accounting for assets in QuickBooks is not easy. The design of the inventory system and the lack of specific features for depreciation makes farm asset accounting in Quick-Books more like riding a bicycle than driving a Cadillac. A future release of QuickBooks or a third-party add-on may someday overcome these shortcomings. Until then, you'll have to make do with what QuickBooks currently offers.

There are solutions to every problem, and this one is no exception. Sometimes all that's necessary is to think about the problem differently. The biggest road-block to farm asset accounting in QuickBooks is a desire to handle all of it *within* QuickBooks. If you're willing to do part of the work outside of QuickBooks—in a spreadsheet program, for instance—you should be able to meet at least basic goals for farm asset accounting.

This chapter talks about simple asset accounting techniques and work-arounds, which operate both within and outside of QuickBooks.

Current Assets and QuickBooks

Current assets are "cash or near cash" assets, including cash, bank accounts, marketable securities, accounts receivable, prepaid expenses, and inventories of grain, market livestock, and supplies. Traditionally, current assets includes things you expect to sell or use up in less than a year.

Another important criteria defining current assets is that you can convert them into cash (sell them off) without reducing the farm's ability to produce output. For example, market hogs are a current asset, because selling them does not cut into the farm's ability to produce pork in the future. But selling part of the sow herd *would* reduce the farm's pork production capacity—breeding livestock are fixed assets, not current assets.

Accounting for most current assets is easy in QuickBooks. Cash and bank account balances are updated automatically when you enter checks, deposits, etc. The same is true of accounts receivable balances, if you use QuickBooks' invoicing and customer accounts features.

Farm inventories, however, are an exception. The design of QuickBooks' inventory system (it was designed for retail businesses) makes it difficult to use for farm inventories such as growing crops, stored grain, and market livestock. This makes generating an accurate farm balance sheet directly from QuickBooks more difficult than it should be.

Here are the basic options for handling farm inventory values in QuickBooks:

- QuickBooks' inventory system *does* work for livestock purchased for resale. See the discussion beginning on page 199. (The inventory system does not work well for *raised* livestock however, because they are not purchased.)

- QuickBooks' inventory system *can be made to work* for inventories of farm production—grain, raised livestock, hay, etc.—if you are willing to use techniques like those described on page 187.

- If you prefer to prepare balance sheets directly from within QuickBooks, you may adjust farm inventory account balances manually as described later in this chapter.

- Currently the most practical way to get farm inventory values into a balance sheet report may be to build the balance sheet outside of QuickBooks. Chapter 12, Essential QuickBooks Reports tells how to use a spreadsheet program for the job.

QuickBooks' Farm Inventory Problems

Why doesn't the QuickBooks inventory system work well for many types of farm inventories? Here are some pieces of the puzzle:

- QuickBooks is designed as an accrual accounting system. A basic concept of accrual accounting is that the cost of buying or producing goods held in inventory (which would include farm production such as grain, hay, or market livestock) is not deducted as an expense until the goods are sold. The costs then show up as a Cost of Goods Sold line on profit and loss reports. But this whole idea is at odds with the cash-basis accounting method used by most farmers, which allows production expenses to be deducted at the time they are paid.

  Profit and loss reports will essentially deduct expenses *twice* if you inventory farm production in QuickBooks in the standard way. How? Besides the usual expense deductions, Cost of Goods Sold will be deducted too!

  Note: *the technique described on page 187 for managing grain inventories offers a work-around for this problem. However, it requires setting up QuickBooks inventory items in a non-standard way.*

- Assigning a value to farm inventories—which you must do if you want them to show up on the farm balance sheet—also determines the amount that will be deducted as Cost of Goods Sold expense when those inventories are sold. When you sell an inventory item, there's no way to prevent QuickBooks from posting the item's value to a Cost of Goods Sold account.

♦ QuickBooks uses the *average costing* method of inventory valuation, which means that the average value of *all* prior purchases or additions to a particular inventory item will be deducted as the cost of that item (as Cost of Goods Sold) when you sell it. This is counter to income tax rules that allow deducting only the *actual* purchase cost of livestock or other goods bought for resale.

Note: *"Keeping Track of Livestock Purchased for Resale", on page 199, offers a work-around for this problem. Setting up a separate QuickBooks inventory item for each group of livestock purchased prevents QuickBooks from averaging the purchase cost of different groups.*

Inventories of supplies pose other problems. For cash-basis tax records, you need to deduct the cost of feed, fuel, and other supplies as expenses. But if you do that, QuickBooks provides no way to also get them into inventory. You could place them in inventory by making manual entries, but that would likely require more time and effort that it would be worth in terms of management information.

Fixed Assets and QuickBooks

Accountants often refer to fixed assets as "plant and equipment". Fixed assets are long-run tools of production, normally owned and used by the business for more than one production cycle. Traditionally, assets you expect to hold for more than one year and those you can depreciate are fixed assets. Examples are machinery, breeding livestock, and land.

QuickBooks also has no special features for keeping track of fixed assets. This is not as big a problem as the farm inventories problem, but it still means that having detailed fixed asset records in QuickBooks requires a lot of manual entries. So if you want *very detailed* records for depreciable and other fixed assets, consider keeping those records outside of QuickBooks.

Here are the basic options for handling fixed asset values in QuickBooks:

♦ If you want to prepare *book value* balance sheets, you need to enter depreciation expense to adjust asset values for depreciation (page 296).

♦ If you want to prepare *market value* balance sheets, you may adjust asset account balances manually as described on page 287.

♦ If you want to maintain asset records at *book value* yet also have the option of preparing *market value* balance sheets, you may want to use the special-purpose adjustment accounts described on page 293.

♦ Currently the most practical way to get fixed asset values into a balance sheet report may be to build the balance sheet outside of QuickBooks. Chapter 12, Essential QuickBooks Reports tells how to use a spreadsheet program for the job.

A **market value balance sheet** is a financial statement which lists assets at their estimated current market value (less selling or disposal costs). Lenders often ask for market value balance sheets, because they show the actual value of the collateral backing up a loan.

A **book value balance sheet** lists assets at "book value"—the remaining value of the owner's investment in them. This remaining value is calculated as the asset's original cost or "basis", minus depreciation. Book value balance sheets are most useful for evaluating farm business profits. They allow estimating a return on investment based on the owner(s) actual investment in the business, excluding consideration of the changing market values of farm assets.

Simple Fixed Asset Accounting for Tax Records

Problem
I just use QuickBooks to keep simple income and expense records for filing income taxes—I don't even use it for preparing balance sheets. Do I need to keep detailed records of depreciable assets and other fixed assets? Do I need to enter depreciation expense?

Solution
No. But you do have to keep track of fixed asset purchases, sales, and disposals for income tax purposes.

Discussion
Every farm business must file income taxes, and that requires records of fixed asset purchases, sales, and disposals. You can accomplish that by setting up just a few fixed asset accounts and posting all fixed asset purchases and sales to them. At year's end you can print a detailed report of fixed asset transactions to take to your tax preparer.

How to Enter Fixed Asset Purchases/Sales/Disposals

This section's focus is on keeping <u>minimal</u> fixed asset records with QuickBooks.

You might set up one Fixed Asset type account and name it:

Fixed Asset Purchases/Sales/Disposals

If you want more detail, set up more accounts. Instead of one main account, you might also add a few subaccounts. Here's a main account and two subaccounts which would provide separate records of fixed asset purchases and sales:

Fixed Asset Changes
Purchases
Sales/Disposals

When you enter a check to record a purchase of machinery, depreciable livestock, land, or some other fixed asset you could use the Fixed Asset Changes:Purchases account as the offsetting account. Likewise when you sell a fixed asset, use the Fixed Asset Changes:Sales/Disposals account when you enter the deposit.

You should also record things like breeding livestock deaths and other fixed asset losses—such as a combine destroyed by fire—in the Fixed Asset Changes:Sales/Disposals account. That assures the lost items will be listed in the report you give to your tax preparer, so he or she can give them proper tax handling.

Because no dollar amount changes hands when a fixed asset is destroyed/lost/dies, you might wonder how to make an entry for the lost item. One way is to enter a deposit to farm checking—as if you sold the item—for a dollar amount of 0 (zero). With an amount of zero, the deposit won't affect the farm checking balance, but the transaction will be listed in the Fixed Asset Changes:Sales/Disposals account.

A more correct approach from an accounting standpoint, would be to make an entry in the Register window of the Fixed Asset Changes:Sales/Disposals account, posting the offsetting amount of the transaction to an equity account. Here's how:

1. **Open the Chart of Accounts window.**

   Choosing Lists|Chart of Accounts is one way to do this.

2. **Highlight the Fixed Asset Changes:Sales/Disposals account by clicking on it.**

3. **Open the account's Register window.**

   Either type Ctrl-R, or choose Activities|Use Register, or click on the Activities button at the bottom of the Chart of Accounts window and then in the pop-up menu click on Use Register.

4. **Add a transaction to the Register for the lost/destroyed item, using zero as the transaction amount.**

The last line of the Register window shown below is an entry for a cow killed by lightning. Be sure the Memo field describes the lost item well, so it will be adequately identified on reports. To avoid confusion it's usually best to use a dollar amount of 0 (zero). Your tax preparer will likely value the lost/destroyed item based on other information, anyway.

Date	Ref	Payee		Decrease	✓	Increase	Balance
	Type	Account	Memo				
02/10/1998	3846	Jackson, Ronnie		5,500.00			-7,455.89
	DEP	Farm Checking [split]	Sprayer and tanks				
02/10/1998	4630	Brown, Randy		1,100.00			-8,555.89
	DEP	Farm Checking [split]	Gelbvieh bull				
11/18/1998		Central Livestock Auction		1,207.07			-9,762.96
	DEP	Farm Checking [split]	Sold 2 cows and heifer				
03/01/1999				0.00			-9,762.96
	TRANSFR	Capital	Cow #1134 died				

Fixed Asset Changes:Sales/Disposals — How Do I?

Record · Edit · Splits · Restore · Q-Report · Go To

Ending balance   -9,762.96
☐ 1-Line
Sort by [Date, Type, Document ▼]

Printing a Report for Your Tax Preparer

At tax time print a separate report for each of the Fixed Asset Changes subaccounts. Filter each report to only include transactions from the desired tax year.

There are several reports you could use, but a QuickReport for each account would serve the purpose well. To open a QuickReport, first highlight the desired account in the Chart of Accounts window, then either type Ctrl-Q, or click on the Reports button at the bottom of the Chart of Accounts window and then in the pop-up menu click on QuickReport.

Accounting for Market Value Balance Sheets

Problem

I want to prepare balance sheets based on current market values of farm assets. What's the best way to assign market values to assets in QuickBooks, before preparing a balance sheet?

Solution

Adjust the asset account balances by making entries in either the General Journal or in each account's Register window. Use an equity account as the offsetting account in these entries (a change in asset values changes the value of the owner's equity in the business).

Discussion

Preparing a market value balance sheet requires, well...market values. Assets need to be listed on the balance sheet at their estimated current market value.

The simplest way to adjust assets to market value in QuickBooks is to make adjusting entries just prior to preparing a balance sheet report, to update asset account balances to current market value. You may do this for accounts representing farm inventories (growing crops, stored grain or produce, market livestock) and fixed assets (machinery, breeding livestock, land). But *never* change the balance of any asset account that carries a specific balance, like a checking account or Accounts Receivable.

If you are comfortable with using a spreadsheet, an alternative is to prepare the balance sheet report in a spreadsheet program. Exporting a QuickBooks balance sheet report to a spreadsheet and adjusting market values there may be simpler than making adjusting entries in QuickBooks. See the Essential QuickBooks Reports chapter, beginning on page 323, for details.

One more thing. One of the requirements of accounting for fixed assets is producing the necessary reports for income tax preparation. Chapter The Essential QuickBooks Reports chapter tells how to print a report which lists fixed asset purchases, sales, and disposals. This is a report you will need to take to your tax preparer.

If you want market value *and* book value balance sheets...

If you want both kinds of balance sheet reports, it's best to maintain book value records and make temporary market value adjustments when you want a market value balance sheet. For market value adjustment examples, see page 302.

No depreciation necessary...

There's no need to make depreciation entries if you only prepare market value balance sheets, unless you want depreciation expense included in the farm's profit and loss reports. Depreciation is only important if you are maintaining asset *book values*, and is irrelevant to the *market value* of assets.

What Asset Accounts Do You Need?

Your first decision about asset accounts in QuickBooks should be deciding how much detail you want in them. The extremes are to have hundreds of accounts—one for each farm asset—or to have just a few accounts representing the combined value of all assets. You probably should be somewhere between those two extremes.

One factor in the decision is how much asset detail you want shown on balance sheet reports. QuickBooks will include as many asset accounts on balance sheets as there are in your chart of accounts. For a balance sheet report to be useful in conveying information to others, such as your lender, it should include enough detail to be informational but not so much as to be an annoyance.

Another factor is where you want to keep track of asset detail, especially for fixed assets—things like their purchase date, current value, etc. Assigning market values is easier if you have a detailed list of farm assets. But that list doesn't have to be maintained within QuickBooks. A handy place to assign market values to an asset list is in a spreadsheet program, because the spreadsheet adds up all of the asset values for you. With this approach you only need a few asset accounts in QuickBooks, each representing a group of assets. An added benefit of using just a few asset accounts is that the job of updating their balances doesn't take long.

Here are examples of asset group accounts you could use. Note that this list includes both current asset accounts and fixed asset accounts:

Growing Crops	Other Current Asset	0
Grain Inventory	Other Current Asset	66,000
Market Livestock Inventory	Other Current Asset	28,000
Chemicals Inventory	Other Current Asset	1,200
Supplies Inventory	Other Current Asset	1,800
Feed Inventory	Other Current Asset	2,400
Breeding Livestock	Fixed Asset	36,000
Machinery	Fixed Asset	124,000
Land & Improvements	Fixed Asset	133,000

You may of course use more accounts and subaccounts for additional detail. But remember, every account you add will show up on balance sheet reports! Keeping the balance sheet simple will let it do a better job of communicating the farm's financial position. So using just a few asset accounts and supplying extra detail as a "supporting attachment" to the balance sheet may be best. You can do that by printing off and including the asset list maintained in your spreadsheet program (or elsewhere).

Shown below is part of a spreadsheet that was used for estimating the current market value of the Grain Inventory account ($66,000) in the list of accounts, above. It is the sort of detail that could be printed as a supplement to the balance sheet report you print from QuickBooks.

	A	B	C	D	E
1	Grain Inventory - Market Value Worksheet				
2-3		Location	Bushels	Value/ Bushel	$ Value / Bin
5	Corn	Bin #1 - South	2000	2.25	$4,500
6		Bin #3 - Northeast	12000	2.25	$27,000
7		ADM	1445	2.25	$3,250
8		Total Corn	15445	bu.	$34,750
9					
10	Soybeans	Bin #2 - Southeast	4200	6.25	$26,250
11		Smith farm bin	800	6.25	$5,000
12		Total Soybeans	5000	bu.	$31,250
13					
14	All Grain		20445	bu.	$66,000

For inventories of farm production, another way to keep track of asset detail is with QuickBooks' inventory system. You can set up Inventory Part items which are linked to a specific asset account. Adjusting the value of these inventory items automatically updates the asset account's balance. See the grain inventory topic on page 187 for details.

How to Adjust Asset Account Balances

Your choices for how to adjust the balances of most asset accounts are to either (1) enter the adjustments in the General Journal, or (2) enter them in a Register window for each account. (There's also a third choice if you use Inventory Part items for farm production. Adjusting the inventory value automatically adjusts the associated asset account balance. See the grain inventory topic on page 187.)

The General Journal lets you adjust all asset balances in a single transaction but requires an understanding of debits and credits. The Register window requires entering a separate transaction for each account, but does not require that you know a debit from a credit (Register windows use the words "increase" and "decrease" instead of "debit" and "credit".)

Suppose the balance in the Grain Inventory account is $66,000 but the current market value of the farm's grain inventory is now estimated at $70,000. The following steps show how to adjust the Grain Inventory balance to the new figure by making an entry in the account register.

Note: *this adjustment technique is only for assets <u>not</u> maintained as Inventory Part items. For Inventory Part items, their value should be adjusted using the inventory quantity and value adjustment activity (Activities|Inventory|Adjust <u>Qty</u>/Value on Hand). And you should never adjust the value of resale livestock inventories, or any similar inventory which must be maintained exactly at cost.*

1. **Open the Chart of Accounts window.**

   Choosing Lists|Chart of Accounts is one way to do this.

2. **Highlight the Grain Inventory account by clicking on it.**

3. **Open the Grain Inventory account's Register window.**

   Either type Ctrl-R, or choose Activities|Use Register, or click on the Activities button at the bottom of the Chart of Accounts window and then in the pop-up menu click on Use Register.

4. **Add a transaction line to the Register to increase the Grain Inventory balance to $70,000.**

   Here's how the register might look when you're done:

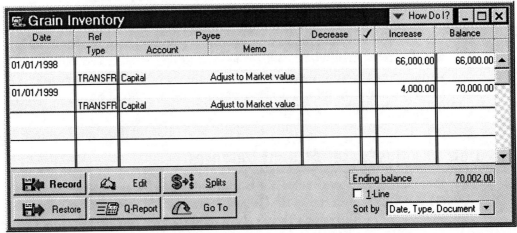

You should always use an equity account (such as Capital or Owner's Equity) as the offsetting account when adjusting accounts to market value, because changes in the value of farm assets affects the value of owner's equity.

Note: *this adjustment technique is only for assets <u>not</u> maintained as Inventory Part items. For Inventory Part items, their value should be adjusted using the inventory quantity and value adjustment activity (Activities | Inventory | Adjust <u>Qty/Value</u> on Hand). And you should never adjust the value of resale livestock inventories, or any similar inventory which must be maintained exactly at cost.*

How to Account for Fixed Asset Purchases and Sales

Producing market value balance sheets by the methods described above is slightly at odds with the fact that you must keep track of purchases, sales, and disposals of *specific* fixed assets for income tax purposes.

To have the necessary tax records you may enter all fixed asset purchases, sales, and disposals in just a few accounts. (This approach was also described earlier in the chapter.) Here are some accounts, with example balances:

Fixed Asset Changes	21,200
Purchases	32,300
Sales/Disposals	-11,100

But here's a problem. The net dollar amount of fixed asset purchases minus sales/disposals ($21,200 in the example above) has *nothing* to do with the market value of the farm's assets.

One way to get around this problem is to make QuickBooks ignore the balance in the Fixed Asset Changes account when you prepare a balance sheet. To do that, *temporarily* enter a transaction in the Register window of the Fixed Asset Changes account to zero out the account balance (by offsetting the entire amount, which is $21,200 in this example, to an equity account). When you prepare the balance sheet report, Fixed Asset Changes won't be included because of its zero balance. Afterwards, return to the Register window of the Fixed Asset Changes account and delete the temporary entry, restoring the correct account balance.

By the way, the Essential QuickBooks Reports chapter, which begins on page 323, tells how to print a report which lists fixed asset purchases, sales, and disposals. This is a report you will need to take to your tax preparer.

Accounting for Asset Book Values: Depreciation & Other Stuff

Problem

What are my options for maintaining farm assets at book value in QuickBooks, so I can prepare book value balance sheets?

Solution

The rest of this section tells about ways to maintain book values in QuickBooks.

Discussion

QuickBooks works better for maintaining assets at book value than at market value. No account balance adjustments are necessary, except in the case of farm inventory accounts (growing crops, stored grain or produce, market livestock).

For fixed assets, you may keep as much or as little detail in QuickBooks as you like. You can either set up an individual account to represent each fixed asset, or set up a few accounts with each one representing an entire category of fixed assets, like machinery or breeding livestock. (For this approach you must maintain the individual asset information outside of QuickBooks, such as in a spreadsheet or a fixed asset management program.)

If you set up individual account's for the farm's fixed assets, accounting for them mostly boils down to three activities: (1) setting up a new account and entering the cost when an asset is purchased, (2) periodically entering depreciation expense, and (3) recording the sale or disposal of the asset, if sold, lost, or destroyed.

By the way, you're on your own when it comes to tracking fixed asset values in QuickBooks. There are no special features for maintaining a depreciation schedule or fixed asset list, and QuickBooks cannot calculate depreciation expense.

Farm inventories (growing crops, stored grain or produce, market livestock—all of which are current assets) are more of a problem. The design of QuickBooks' inventory system provides no easy way to maintain correct values for inventories of farm production, other than to adjust them manually. If you want farm inventories listed on the balance sheet at book value, you will have to adjust the inventory accounts' balances before preparing the balance sheet. Making the adjusting entries is not hard, but calculating the book values of those accounts certainly is!

One more thing. One of the requirements of fixed asset accounting is being able to produce the necessary reports for preparing income taxes. The Essential QuickBooks Reports chapter tells how to print a report which lists fixed asset purchases, sales, and disposals. This is one of the reports you will need to take to your tax preparer.

You can generate market value balance sheets too!

An advantage of maintaining assets at book value in QuickBooks is that generating a market value balance sheet is also possible. You can get book value balance sheet reports directly from QuickBooks, and get market value reports by exporting a balance sheet report to a spreadsheet, then changing the asset account balances there to their estimated market values. See the Essential QuickBooks Reports chapter, beginning on page 323, for details.

What Asset Accounts Do You Need?

As mentioned above, you can either maintain asset values in individual accounts representing each asset, or by using accounts which represent asset groups.

Using just a few accounts to represent broad asset groups is best for assets with balances that will need adjustment from time to time. A small number of accounts means fewer adjusting entries will be necessary. Because farm inventory balances will need adjusting to approximate book values, they are good candidates for being represented by group accounts, like these:

Growing Crops	Other Current Asset
Grain Inventory	Other Current Asset
Market Livestock Inventory	Other Current Asset
Chemicals Inventory	Other Current Asset
Supplies Inventory	Other Current Asset
Feed Inventory	Other Current Asset

Asset group accounts were discussed in detail earlier in this chapter.

Fixed asset account balances don't need adjustment if you use them to track book values, so either asset group accounts or individual accounts for each fixed asset will work.

Setting up individual fixed asset accounts lets you keep track of the original cost (basis) and depreciation of each asset within QuickBooks, but may add too more detail to balance sheet reports than you want. Here's an example of how individual fixed asset accounts may be arranged:

Breeding Livestock
1998 Angus Cows 11
Basis
Accum. Depr.

> 1999 Gelb/AngXCows 23
>> Basis
>> Accum. Depr.
> Machinery
>> JD 7700 '97
>>> Basis
>>> Accum. Depr.
>> Kinze Planter '96
>>> Basis
>>> Accum. Depr.
>> Case-IH Disk '94
>>> Basis
>>> Accum. Depr.
> Buildings
>> Machine Shed '95
>>> Basis
>>> Accum. Depr.
>> Grain Leg & Pit '94
>>> Basis
>>> Accum. Depr.
> Land
>> Home farm, 220 acres
>> Smith place, 120 acres

The Basis and Accum. Depr. subaccounts allow keeping track of each assets' original purchase cost and accumulated depreciation amount. They also cause the parent account's balance (in the Chart of Accounts window) to always reflect the asset's remaining book value, because a parent account's balance is always the net total of its subaccounts.

It is important to note that if you want this much fixed asset detail, you will need to set up a new asset account and necessary subaccounts for *each* asset that is purchased.

Tracking fixed assets by...Payee Name?!?

The downside of using individual fixed asset accounts is that they will all show up on balance sheet reports, and that may be more detail than you want.

A possible alternative (depending on your version of QuickBooks) is to set up a smaller number of accounts representing asset groups, and track individual assets within each account by typing the asset's name in the Payee Name field of the transactions. You can then get a report, filtered to find a specific Payee Name, to see all transactions pertaining to a particular asset. See the QuickBooks User's Guide or Help system for details about this technique.

Accounts for Book and Market Values

Most people who maintain assets at book value and prepare book value balance sheets also want to prepare market value balance sheets part of the time.

One way to update assets maintained at book value to their current market values is by using special-purpose subaccounts for making the market value adjustments. Market value adjustment subaccounts will work with either group or individual asset accounts, but the basic idea is probably easiest to understand for an individual asset account:

> Machinery
>
> ...
>
> JD 7700 '97
>   Basis
>   Accum. Depr.
>   Mkt Value Adjustment

Note that the accounts for this tractor asset are identical to the ones shown earlier, but with the addition of a Mkt Value Adjustment subaccount.

How is the subaccount used? Because it will normally have a zero balance, the Basis and Accum. Depr. subaccounts will usually control the parent account's balance—which will therefore be at book value. To prepare a market value balance sheet, a *temporary* entry is made in the Mkt Value Adjustment subaccount to adjust the parent account's balance to current market value. After preparing the balance sheet the temporary entry is deleted, which returns the parent account's balance to book value. A complete example is presented later in this chapter.

How to Enter an Asset Purchase (with No Trade-In)

To enter a fixed asset purchase in QuickBooks, first you need to assure that the necessary fixed asset account(s) are present, as described above. Then enter the purchase itself—usually just a simple transaction involving a check or other form of payment.

Here are the Fixed Asset type accounts you might need for a new hay rake:

> Machinery
>
> ...
>
> Hay rake'99

Basis
Accum. Depr.

And here's how you might enter a check to record purchase of the rake:

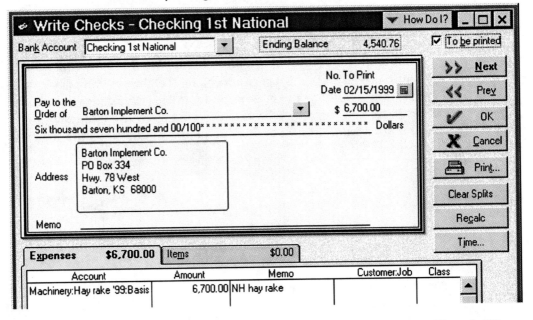

Notice that the account selected in this transaction is Machinery:Hay rake'99: Basis. This transaction records the original basis, or purchase price, of the hay rake.

One more thought about fixed asset purchases: let's clear up any confusion about "expensing" fixed assets. The Federal tax code currently allows expensing part or all of certain fixed asset purchases (the Section 179 Deduction). But just because your tax preparer may expense an asset when filing your tax return, that doesn't mean you should consider the asset an expense in the farm accounting records. Regardless of their tax handling, all fixed assets should be entered in the farm accounts and depreciated.

What about purchases of farm inventory assets?

If you maintain any farm assets as QuickBooks inventories, such as resale live-stock (page 199) or stored grain and other farm production (page 187), those inventories should be entered at (or adjusted to) book value if you are keeping book value asset records.

Assets *not* maintained as QuickBooks inventories, but which are considered farm inventories on the balance sheet are either (1) not normally purchased (growing crops, stored grain or produce, market livestock), or (2) are usually entered as expenses when purchased (feed, supplies, fuel, seed, etc.). These have no way to get included in the balance sheet except by account balance adjustments. The adjustments discussed later in this section assign "estimated" book values to those assets.

How to Enter an Asset Purchase with a Trade-In

When you trade in one fixed asset as part of the purchase of another, the remaining book value of the trade-in becomes part of the book value of (or, basis in) the new asset. The transaction which records purchase of the new asset needs to also transfer the old asset's remaining book value to the new one.

Suppose the same hay rake purchase described above had instead included trading in an old hay rake, along with a cash payment of $5,000. Let's say the old rake was represented in the chart of accounts by these Fixed Asset accounts and account balances:

Machinery	X,XXX
...	
JD rake'97	2,000
Basis	3,500
Accum. Depr.	1,500

The thing you need to know before entering the new rake purchase, is the remaining book value of the old rake—the trade-in. As you can see from the account balance shown above, the old rake's remaining book value is $2,000. With this information, you can go ahead with entering the $5,000 check for the new rake:

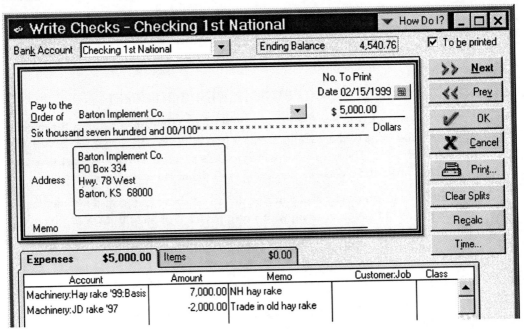

The first detail line of the transaction posts $7,000 to the new hay rake's Basis subaccount: Machinery:Hay rake'99:Basis. The second detail line posts -$2,000 to Machinery:JD rake'97. The net amount of the check is the correct amount—$5,000—and the new hay rake's account receives the full $7,000 basis (the $2,000 book value of the trade-in, plus the $5,000 payment). Also, the book value remaining in the old hay rake's account is fully offset (zeroed out) by the -$2,000 amount of the second detail line.

What if a trade-in has no remaining book value? It's still a good idea to list the trade-in (at a price of zero) in the purchase transaction even though it won't affect the basis of the new asset, because:

- The transaction which records the asset purchase also records the fact that the old asset was traded in, and to whom, and for which new asset.

- The trade-in will appear in reports of the year's asset purchases/sales/disposals which you give to your tax preparer, allowing proper tax handling.

How and When to Enter Depreciation Expense

Depreciation is a way to "expense" the cost of an asset over a period of several years, according to a predefined plan (a depreciation method) of your choice. Entering depreciation expense in the farm records accomplishes two things. It: (1) records the depreciation as an expense in a particular accounting year (depreciation expense will appear on profit and loss reports for that year), and (2) serves as an offset or reduction in the remaining book value of the farm's depreciable assets.

Most farm record keepers enter depreciation expense once per year, usually some time after the accounting year has ended and they've found time to get the necessary information together. Those who use the depreciation amounts from their income tax return must wait until they get those figures back from their tax preparer.

That brings up a question: which depreciation method should you use? Using the depreciation amounts calculated for tax purposes is easy—you don't have to calculate depreciation on your own. But most tax depreciation methods are "accelerated" methods, meaning they expense assets faster than the assets actually decline in value or usefulness. Other depreciation methods such as straight line or double declining balance put the burden of calculating depreciation on you, but offer more rational approaches to depreciating farm assets. And, the calculations really aren't difficult.

Ask your accountant, farm management advisor, or lender which depreciation methods they would recommend for your farm business.

Note: some lenders have settled on income tax depreciation expenses as their standard for analyzing farm records. If that's true of your lender, you may want to use depreciation figures from the farm's tax return for the sake of conformity.

When very accurate book values are the goal, a **partial year's depreciation expense** may be entered. For example, to prepare an accurate book value balance sheet as of the first day of July you might enter one-half of the expected annual depreciation expense. The remainder of the depreciation expense would be entered later, to complete the year's records.

Few farm businesses face the need to produce book value balance sheets with that much accuracy. However, larger operations—particularly those with externally-audited financial statements—may need to do so from time to time.

One thing you'll need before you enter depreciation expense is an account where the depreciation expense can be posted. If you don't already have one, set up a Depreciation Expense account of the Expense type.

If you use a detailed set of fixed asset accounts like the ones shown earlier (with an individual account representing each asset, and separate Basis and Accum. Depr. subaccounts), you'll have a lot of depreciation expense entries to make. You could enter them one at a time, in each Accum. Depr. subaccount's Register window. But a faster way is to enter them all in a single General Journal entry. Here's a short example, with entries for just three assets:

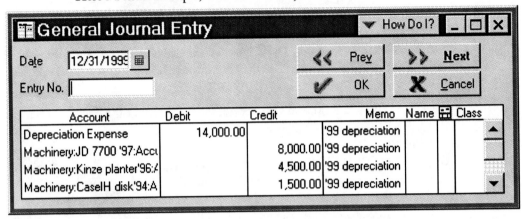

This transaction is dated 12/31/99 but likely would be entered sometime during the year 2000. (Depreciation expenses for a particular year often aren't calculated until after completing the accounting year.) Be sure to date depreciation expense transactions so that they fall within the desired year—if you don't, they won't be included in that year's profit and loss reports.

As for the effects of the transaction, the first line debits Depreciation Expense, which adds to the balance of that expense account. The Account column is too narrow to show the complete account names, but the next three lines each credit a machinery account's Accum. Depr. subaccount. That increases the accumulated depreciation balance in those accounts, and thus reduces the remaining book value of the three assets.

Finally, if all this debit and credit stuff is confusing, you can always fall back to "Plan B"...which is to make individual depreciation expense entries in the Register window of each asset's Accum. Depr. subaccount.

Crediting an asset account decreases its balance. **Accumulated depreciation accounts are almost always credited,** which causes them to normally carry a negative balance. That's why they are sometimes referred to as "negative asset" or "contra-asset" accounts.

How to Enter Asset Sales/Disposals

Usually the selling price of a fixed asset will be more or less than its remaining book value. The entry recording the sale needs to remove (offset) the remaining book value of the asset, and also post the difference between the selling price and the book value to a capital gains and losses account (an account you use for accumulating capital gains and losses).

The account you set up for recording capital gains and losses should normally be of the Other Income type. Other Income accounts appear toward the bottom of profit and loss reports, which allows considering capital gains and losses apart from regular farm income. Here's the account we'll use in the example below:

Capital Gain/Loss   Other Income

Capital gains or losses figured on your income tax return and those you enter in QuickBooks involve the same basic idea. However, the numbers are unrelated and likely to be different, due to differences in calculation methods, depreciation amounts, and other factors. There is no need for your capital gain and loss entries to match those on the farm's income tax return.

Now let's look at a fixed asset sale example. Suppose the farm owns a bull with a remaining book value of $400, as shown by these accounts:

Breeding Livestock	Fixed Asset	X,XXX.XX
...		
Angus Bull '97	Fixed Asset	400.00
Basis	Fixed Asset	1,800.00
Accum. Depr.	Fixed Asset	1,400.00

The bull is sold to J.T. Smith for $1,100 ($700 more than the remaining book value). Here is a deposit entry which records the sale, removes the remaining book value from the fixed asset account, and records the $700 gain:

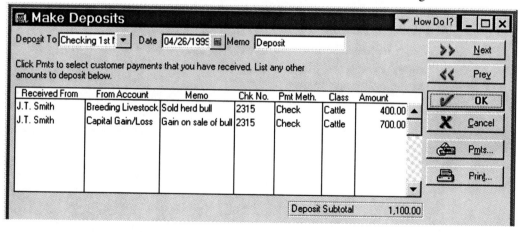

The From Account column isn't wide enough to show the complete account name on the first detail line, which is Breeding Livestock:Angus Bull'97. Note that this is the *parent* account representing the bull, not the Basis or Accum. Depr. subaccounts. This keeps the bull's basis (purchase price) and depreciation information intact for reference purposes, yet offsets the bull's remaining book value of $400—the net balance of the Angus Bull'97 account will be $0 (zero) after the deposit is entered.

Some accountants may object to the idea of making fixed asset transfers in a deposit entry and of "splitting up" a deposited check amount as shown in this example. But the positives of this approach outweigh the negatives. Making all the necessary entries as part of a deposit is easier and more familiar to most users than the alternative, which is to make the entries in the General Journal.

From time to time you'll also need to record things like death losses of breeding livestock or other fixed asset losses—maybe a piece of machinery destroyed by fire. A good place to make such entries is in the Register window of the appropriate fixed asset account (or, you may use the General Journal if you prefer).

If the bull described in the previous example had died, here's how you might enter the loss in the Register window:

1. **Open the Chart of Accounts window.**

   Choosing Lists | Chart of Accounts is one way to do this.

2. **Highlight the Breeding Livestock:Angus Bull '97 account by clicking on it.**

3. **Open the account's Register window.**

   Either type Ctrl-R, or choose Activities | Use Register, or click on the Activities button at the bottom of the Chart of Accounts window and then in the pop-up menu click on Use Register.

4. **Add a transaction line to the Register to zero out the remaining book value, posting the offset amount to the account you use for recording capital gains and losses:**

Date	Ref	Payee		Decrease	✓	Increase	Balance
	Type	Account	Memo				
04/01/1997	4728	Windale Cattle Co.				1,800.00	1,800.00
	CHK	Checking 1st National	Angus bull, tag # 5-331				
12/31/1997				700.00			1,100.00
	GENJRNL	Depreciation Expense	'97 depreciation				
12/31/1997				700.00			400.00
	GENJRNL	Depreciation Expense	'98 depreciation				
04/01/1999				400.00			0.00
	GENJRNL	Capital Gain/Loss	Bull died 3/27/99				

Breeding Livestock:Angus Bull '97 — How Do I?

Record   Edit   Splits

Restore   Q-Report   Go To

Ending balance   0.00

☐ 1-Line

Sort by   Date, Type, Document

The last line of this Register window records the death of the bull and de-creases the account's remaining book value to zero, by transferring the remaining basis in the bull to the Capital Gain/Loss account.

What about sales and disposals of farm inventory assets?

Sales of assets that are considered farm inventories (stored grain or produce, market livestock, etc.) are normally entered as farm income. Entering sales of items maintained as QuickBooks inventories (as Inventory Part items) will automatically adjust inventory account balances. For inventory assets *not* maintained as QuickBooks inventories, account balance adjustments are necessary to allow for inventory decreases resulting from sales.

Adjusting Farm Inventory Accounts to "Book" Value

For farm businesses using QuickBooks, the main "bump in the road" that leads to a book value balance sheet is the problem of valuing inventories. Farm inventories are current assets such as growing crops, stored grain or produce, market livestock, supplies, fuel, seed, and so on.

The design of QuickBooks' inventory system provides no automatic way to maintain correct values for farm inventories. To have inventories listed on the balance sheet at (approximately) their book value, you must adjust inventory values (for items maintained as QuickBooks inventories) or adjust the balances of inventory accounts (for assets not maintained as QuickBooks inventories) before preparing the balance sheet. ("Accounting for Market Value Balance Sheets", on page 285, talks about inventory account balance adjustments. Refer to that section if you need detailed instructions—they won't be repeated here.)

Accounting 101:   Not true book values...

The book value of an inventory should be calculated from the actual costs of producing or purchasing it. Accounting purists will frown on the suggestion that you "adjust" book values of farm inventories. But that approach is a practical compromise between the limitations of Quick-Books and the goal of producing book value balance sheets with moderate ease.

The big problem in adjusting inventories to book value is determining what book values are appropriate! To approximate a true book value, inventory account balances should be adjusted to the lesser of (1) their estimated production or purchase cost, or (2) their current market value. Accountants know this as "lower of cost or market" valuation.

Here's a simple example. Suppose you are preparing an end-of-year book value balance sheet. Anhydrous ammonia fertilizer was purchased and applied to 350 acres of land in the fall, to be planted to corn in the spring. The cost of the anhydrous ammonia should be included in the balance of the Growing Crops inventory account. If you were preparing a book value balance sheet at mid year, the Growing Crops balance would be adjusted to include all costs invested in the corn crop—seed, fertilizer, insecticide, fuel, labor, etc.

Estimating book values accurately is not easy. You'll probably need to strike a reasonable compromise between accuracy and the amount of effort necessary to attain it. One approach is to use production cost estimates—either your own, or those published by a university or other source—to arrive at reasonable inventory values. Maybe more important, talk to the person or people who will be using the book value balance sheet you prepare—whether that's a lender, a farm management advisor, or a partner. Get their opinions on how inventory book values should be estimated.

One other thing you should do is document the methods and information you use. Write down how inventory book values were estimated and include it along with the balance sheet report. (Balance sheet reports are discussed in chapter 12.)

Adjusting Book Value Records to Market Value

Most people who maintain assets at book value also want to be able to prepare market value balance sheets part of the time. One of the alternatives for this is to prepare the market value balance sheet with a spreadsheet program, as described in chapter 12. But another alternative is to use special-purpose subaccounts for adjusting asset account balances to market value, then preparing the market value balance sheet report directly from within QuickBooks.

The account setup needed for this technique was described on page 293. Let's look at a simple example of how to make the market value adjustment—we'll do it for a single asset.

Suppose a tractor is represented in the chart of accounts by these fixed asset accounts and account balances:

Machinery	X,XXX
...	
JD 7700 '97	48,000
Basis	72,000
Accum. Depr.	-24,000
Mkt Value Adjustment	0

The tractor has a book value of $48,000 as the account balances show. Before preparing a market value balance sheet, an entry can be made in the Mkt Value Adjustment subaccount to adjust the parent account's balance to estimated market value—which we'll assume is $53,000 for this example.

The entry could be made in the subaccount's Register window or in the General Journal. You would normally be adjusting many asset accounts to market value at the same time, which could be done more quickly in the General Journal because it allows debiting or crediting multiple accounts in a single entry. Here are the steps involved:

1. **Open the General Journal window.**

   Choose Activities|Make Journal Entry.

2. **Enter the account balance adjustment.**

   The purpose of the entry is to increase the tractor account's balance by $5,000, which can be done by debiting (increasing) the Mkt Value Adjustment subaccount for $5,000 and crediting (increasing) Owner's Equity by the same amount. (An equity account is used as the offsetting account for asset value adjustments, because a change in the value of the farm's assets affects the value of the owner's investment in the farm business.)

   Here's an example of the entry:

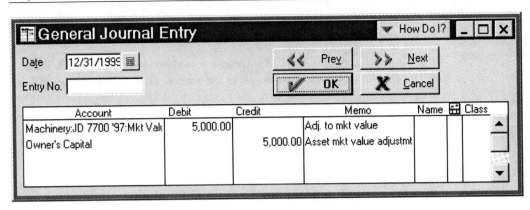

Remember, this example only adjusts the balance of one asset account, but in an actual situation it would more likely be used to adjust the balances of many accounts.

3. **Click OK to save the entry and close the General Journal window.**

After making the entry, the account balances would look like the following. Note that the parent account's balance is now $53,000—an increase of $5,000 over its book value:

Machinery	X,XXX
...	
JD 7700 '97	53,000
Basis	72,000
Accum. Depr.	-24,000
Mkt Value Adjustment	5,000

With the adjustment made, at this point a market value balance sheet could be prepared—assuming similar entries had been made for all asset accounts needing adjustment to market value.

The illustration on page 335 shows an example of Mkt Value Adjustment subaccounts in a balance sheet report.

After printing the market value balance sheet, the account balances would need to be returned to book value before preparing a book value balance sheet. The two options for doing this are to either (1) delete the adjustment entry, or (2) post-date the adjustment entry.

Deleting the adjustment entry

There are different ways to delete the account balance adjustment entry, but using the General Journal would be easy:

1. **Open the General Journal window.**

Choose Activities|Make Journal Entry.

2. **Find the entry in the General Journal window.**

Click on the Prev (previous) button until you find the entry. Depending on how the original entry was dated, you may have to click the Prev button many times to locate it.

Another option for locating the entry would be to use QuickBooks' Find command.

3. **Delete the entry.**

Either type Ctrl-D, or choose Edit|Delete General Journal.

4. **Click OK to save the entry and close the General Journal window.**

The adjusting entry should now be gone and the tractor account's balance returned to book value.

Post-dating the adjustment entry

This approach does not delete the adjustment entry but only moves it into a future accounting period. The entry will then be available to reuse some time later, when preparing another market value balance sheet. Here are the steps involved:

1. **Locate the adjustment entry in the General Journal window.**

Follow steps 1 and 2 as described above, under "Deleting the adjustment entry".

2. **Change the Date field's entry to a future date.**

Date the entry so that it falls within a *future* accounting period. The original journal entry shown above as dated 12/31/1999, which implies it was used for preparing a market value balance sheet as of the end of 1999. Changing the date to 12/31/2002 or 12/31/2010, etc., would exclude the entry from balance sheets prepared for the current accounting period.

3. **Click OK to save the entry and close the General Journal window.**

The benefit of this approach is that you don't have to completely re-enter the adjusting entry every time you want to prepare a market value balance sheet. Here's how the adjustment entry could be reused at a later date, such as to prepare a market value balance sheet at the end of 2000:

1. **Open the General Journal window.**

Choose Activities|Make Journal Entry.

2. **Find the entry in the General Journal window.**

Click on the Prev (previous) button until you find the entry. Depending on how the original entry was dated, you may have to click the Prev button many times to locate it.

Another option for locating the entry would be to use QuickBooks' Find command.

3. **Change the date of the entry to a date within the desired accounting period.**

Changing the date to 12/31/2000 would allow the entry to adjust account balances for a balance sheet prepared at the end of the year 2000.

Edit the entry as necessary so that it correctly updates all of the necessary asset balances to market value.

Besides changing the market value adjustment amounts on existing lines of the entry, you may need to add lines for assets accounts that have been added since the last time you used the entry.

1. **Click OK to save the entry and close the General Journal window.**

2. **Prepare the market value balance sheet.**

After printing the balance sheet, the adjustment entry could again be post-dated to a future date, to remove it from the current accounting period.

CHAPTER 11

Farm Loan Basics

Q uickBooks basically separates liabilities into two types: current liabilities and long-term liabilities.

A *current liability* is any debt which must be paid within a year. QuickBooks has three different current liability account types: (1) Accounts Payable, which are the typical monthly bills you need to pay to the farm supply dealer, veterinarian, repair shop, electric company, etc., (2) Credit Cards, and (3) Other Current Liabilities, which are typically operating notes and short-term lines of credit.

QuickBooks provides special handling for the Accounts Payable and Credit Card types. The Accounts Payable account (your chart of accounts should have only one of these) is managed by the Enter Bills and Pay Bills activities. As for Credit Card type accounts, you may have any number of those—one for each credit card you use. The Enter Credit Card Charges and Reconcile activities let you manage credit card accounts. Working with Accounts Payable and Credit Card accounts is discussed in chapter 5, Farm Expenses & Payables.

Long-term liabilities are debts to be paid off over more than one year. Any loan with a repayment period of more than one year is a long-term liability. Typical examples are loans for machinery, breeding livestock, and land.

This chapter focuses on accounting for loans identified by the Other Current Liability and Long Term Liability account types.

Setting Up Accounts for Loan Liabilities

Problem
What's the best way to represent loans in the QuickBooks chart of accounts?

Solution
Set up an individual account for each loan.

Discussion
Using an individual QuickBooks account for each loan provides a lot of advantages. Here are some of them:

- The account name can identify the original principle amount of the loan.

- A quick glance at the Chart of Accounts window shows the loan's unpaid principle balance.

- Detail on all payment transactions for a particular loan is available in the loan's Register window or by printing a QuickReport for the account.

How to Set Up a Loan Account

You should set up a new account each time the farm business borrows funds on a new loan...with one exception. Does the farm business borrow funds on a line of credit (also known as a master note) that is basically the same loan, renewed annually—possibly with a different maximum amount each year? If so, there's no need to set up a new account each time the line of credit is renewed. You may re-use the same account.

To set up a new loan account:

1. **Open the Chart of Accounts window.**

   Choosing Lists|Chart of Accounts is one way to do this.

2. **Open the New Account dialog.**

   Either type Ctrl-N, or click on the Account button in the lower part of the Chart of Accounts window and then select New from the pop-up menu.

3. **Fill in the loan account information.**

   Here's an example, for an operating note:

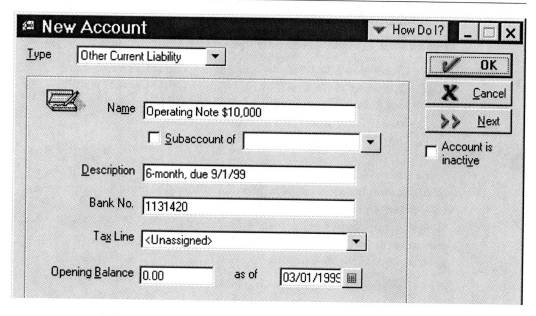

Here are comments on some of the important fields:

Type
Choose Other Current Liability for loans due in less than a year. Choose Long Term Liability for loans to be paid over more than one year's time.

Name
What you enter in this field is what will appear in the Chart of Accounts window, so make the name as descriptive as possible. Including the loan's original principal amount as part of the name is a good idea—it helps identify the loan and, in the case of a line of credit or master note, helps you remember the maximum amount that can be borrowed.

Description
Use this field for anything you want, such as loan terms and/or the due date.

Bank No.
If you know the number the bank has assigned to the loan, you can enter it here for reference purposes, or use this field for any other information.

Tax Line
Normally, no tax line is assigned to loan accounts.

Opening Balance
The opening balance should *always* be zero for new loans. (The transaction which records borrowing of the loan funds will supply the initial account balance.) The only time an opening balance should be included is when setting up already-existing loans in a new QuickBooks company file.

Click OK to close the New Account dialog.

Here are some examples of accounts representing loans:

Credit Line $32,000	Other Current Liability
Operating Note $10,000	Other Current Liability
Hay Rake Note, 6-Month $5,000	Other Current Liability
Tractor Note-AgriMech $55,000	Long Term Liability
Land Note-1st America $120,000	Long Term Liability

Some people like to separate **intermediate liabilities** and **long-term liabilities** on balance sheet reports, but QuickBooks does not automatically make that distinction. If you want intermediate and long-term liabilities to be grouped separately on balance sheets, set up separate parent accounts for both and enter the individual liabilities (loans) as subaccounts of the two parent accounts.

Intermediate Liabilities
    Tractor Note-AgriMech $55,000
Long-Term Liabilities
    Land Note-1st America $120,000

Entering Loan Proceeds (Receiving Borrowed Funds)

Problem

How should I enter the borrowed funds which I receive them?

Solution

Normally they are entered as a deposit.

Discussion

Borrowed funds which flow through the farm checking account are entered like any other deposit. When the funds don't flow through farm checking you use a slightly different approach.

Keeping farm and personal loans separate

Deciding whether a particular loan is for farm or personal use is a frequent problem in sole proprietorships.

Suppose you borrow $15,000 as a farm operating loan, and the next day pay $15,000 "to boot" to trade your old speed boat for a new one.

If separating your farm and personal financial records is difficult, the IRS could make the case that the $15,000 farm operating loan was really a boat loan and—unless you're a fish farmer—disallow deducting the loan interest as a farm expense.

To prevent future tax audit problems, keep farm and personal records separate. Either keep them in separate QuickBooks company files or use the other methods described in this book. A clear record of all owner withdrawals from the farm business and a good "paper trail" showing the details of farm transactions is the best way to prevent future problems.

Receiving Loan Funds to the Farm Checking Account

When the bank that issues the loan is also the bank where the farm checking account is located, typically the borrowed funds will be deposited directly to the checking account. After setting up a liability account for the loan, all you need to do is make a corresponding deposit entry in QuickBooks. The same is true if the loan is from someone else and you actually have a check to deposit.

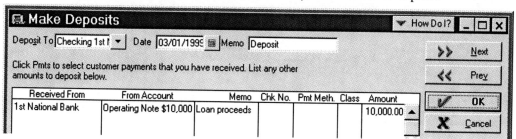

You also enter a deposit each time funds are advanced on a line of credit or master note:

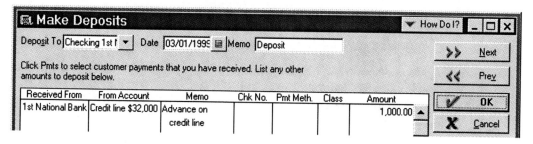

Entering Loans when Funds are Not Deposited to Farm Checking

Borrowed funds aren't always deposited in a bank account. Typical examples are dealer financing of machinery or input purchases, or owner financing of land. In such cases there are no funds to deposit—you just need to set up a new Quick-Books account to represent the loan and then enter the liability balance in it.

One way to do this is with a General Journal entry:

1. **Set up the new loan account, as described earlier in this chapter.**

   The name of the loan account used in this example is "Tractor Note-AgriMech $55,000". It represents $55,000 borrowed from AgriMech Finance (a machinery dealer's financing source) to purchase a tractor.

2. **Choose Activities|Make Journal Entry.**

   The General Journal window will open.

3. **Enter the loan and other information in the General Journal.**

   To keep the example simple, first let's consider what would happen if the tractor's entire purchase price was paid by the $55,000 loan—no cash payment, or trade-in, or other complications.

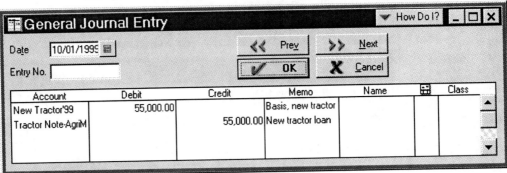

As with any double entry accounting system *every* QuickBooks transaction posts to at least two accounts and always keeps debits and credits in balance. (Usually this happens "behind the scenes" so you may not realize it's happening). Here the loan amount is exactly offset by posting to New Tractor'99, a fixed asset account representing the new tractor. After clicking OK, both the loan account and the tractor asset account will have a balance of $55,000.

The plot thickens when we add "real-world" features to the transaction. Shown below is a similar General Journal entry, but this time for a $68,000 new tractor. The transaction includes a $10,000 cash payment and a trade-in (an older tractor with $3,000 remaining book value) in addition to the $55,000 loan:

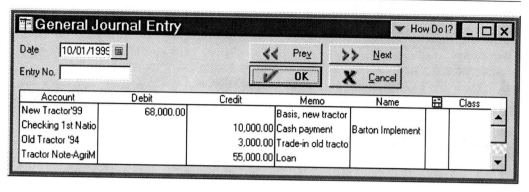

As before, debits equal credits. The $68,000 amount posted to New Tractor'99 exactly offsets the total of amounts posted to the other accounts.

You may wonder about the $10,000 payment drawn from the farm checking account. The actual check written to the machinery dealer would have had a check number and other check features. But the General Journal doesn't provide spaces for entering all that. To see how this entry is recorded in the farm checking account, open a Register window for that account. (Highlight the account in the Chart of Accounts window by clicking on it, then type Ctrl-R.) Here's an example:

Date	Number	Payee		Payment	✓	Deposit	Balance
	Type	Account	Memo				
09/26/1999						1,100.00	38,571.55
	DEP	-split-	Deposit				
10/01/1999						458.45	39,030.00
	DEP	Capital Retain Certificat	Deposit				
10/01/1999		Barton Implement Co.		10,000.00			29,030.00
	GENJRNL	New Tractor'99 [split]	Cash payment				

As you can see, everything is there except a check number. But what if you wanted to print a check for the $10,000 payment? There is no To be printed check box in the Register window, so you might consider opening the transaction in the Write Checks window, so you can print it as a check. But if you click the Register window's Edit button you'll find that you're returned to the General Journal entry, not the Write Checks window. There's no way to open the entry as a check, so that it can be printed.

So why not enter the transaction as a check instead, and bypass all that confusing debits and credits stuff? You can certainly do that. Here is the same transaction entered as a check instead of as a General Journal entry (it accomplishes the same result):

313

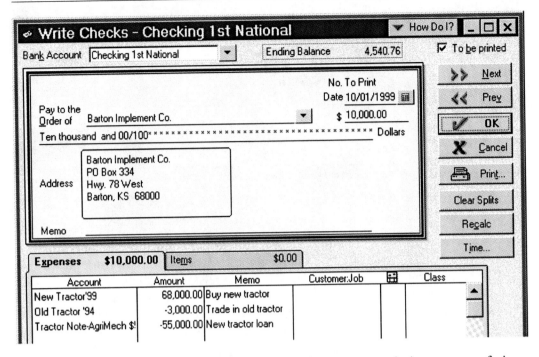

This negative dollar amounts in this check entry may make it seem as confusing as the General Journal entry shown earlier—but at least you could print this one as a check! The transaction increases the balance of the New Tractor'99 account by $68,000. It removes the $3,000 remaining book value of the trade-in, Old Tractor'94. And it increases the balance on the loan account, Tractor Note-AgriMech $55,000, by the $55,000 amount of the loan. The $10,000 check amount is simply the net difference of the detail lines ($68,000 - $3,000 - $55,000 = $10,000).

Entering Loans with Additional Costs or Fees

Some loans have addition fees assessed up front, as part of the loan transaction. Examples are title fees, closing costs, and loan processing fees.

Assuming a $1,000 loan processing fee charged on a $50,000 loan, consider these possibilities:

♦ If you pay the $1,000 fee directly by check, enter the loan and the check separately, with the loan entered as a deposit to farm checking.

♦ If the lender withholds the $1,000 fee from the loan proceeds so that only $49,000 is left to deposit, the effect on your bank account is the same as if you had written a check for the fees. You could enter the $50,000 loan and the $1,000 check separately, as described above. But if you want your records to show that a net amount of $49,000 was deposited, you could record both the loan and the fee in a deposit:

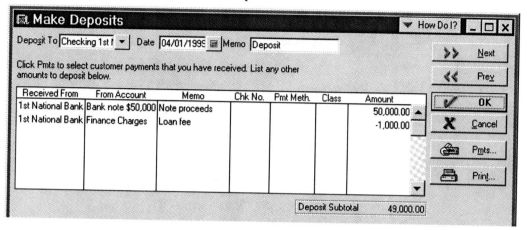

- If the fees are added as an increase in the loan principle to $51,000, but the loan proceeds—the funds you actually receive—are still $50,000, enter the deposit this way:

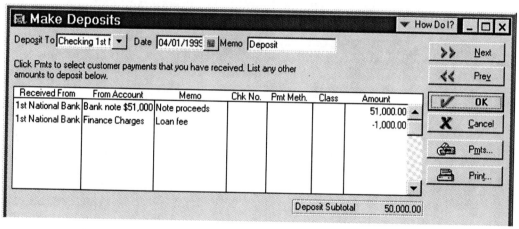

The loan principle amount is now $51,000. But the $1,000 deduction for the loan processing fee reduces the deposit to the actual amount of funds received, $50,000.

315

Entering Loan Payments

| **Problem** | How should I enter loan payments, to properly reduce the outstanding principle balance in my records and record the interest expense? |

| **Solution** | Normally, you will enter the payment as a check or a bill. |

| **Discussion** | Paying a loan is simple. Just be sure to separate the principle payment amount and interest expense by entering them on separate lines of the check entry. |

Paying a Loan Directly, by Check

Here's an example loan payment on an operating loan. The payment retires the entire principle amount ($10,000) and records interest expense of $555.

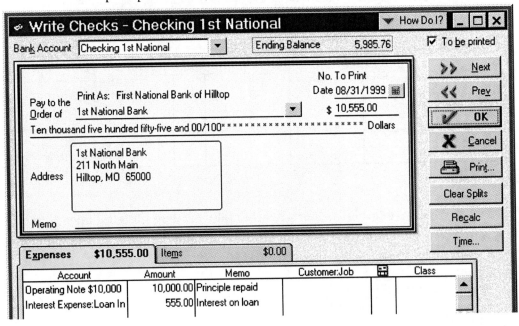

What if the payment is for interest only? Or for principle only? Here's an example check written to pay down part of the principle outstanding on a line of credit.

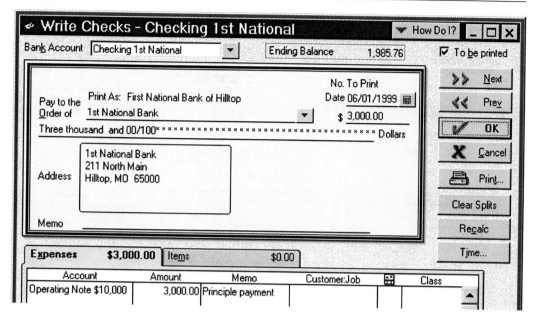

Loan payment entries are basically the same for any kind of loan. One minor exception is that the interest paid on mortgage loans should be posted to a mortgage interest account, separate from other interest expense. Mortgage interest expense is reported separate from other interest on Federal tax forms.

Paying a Loan as a Bill

If you want to be reminded of an upcoming loan payment's due date, consider entering the payment in advance of the due date, as a bill. Due dates and payment amounts are displayed in the Reminders window if you have the Reminders feature turned on (page 28), which should help assure the payment gets made on time.

Here's an example land loan payment entered as a bill, on the Bills form:

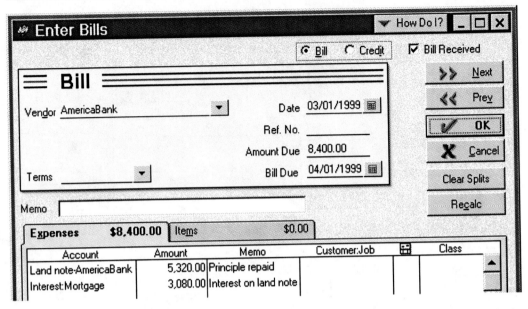

Note that the bill was entered on 3/1/99, but the payment due date is 4/1/99. As the due date approaches, this bill should be listed in the Reminders window:

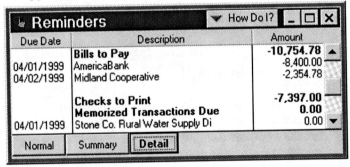

To actually issue the loan payment, select this bill in the Pay Bills window (Activities | Pay Bills), then print or handwrite the check. For bill payment examples see chapter 5, the Farm Expenses & Payables.

Entering a Loan Payment When You Don't Know the Interest and Principle Portions

Have you ever needed to record a loan payment without knowing what part of the payment was principle and what part was interest? This is often a problem on level-payment loans. You pay the same payment amount each time, but the portions that are principle and interest change with each payment.

Here's one way to handle the problem. If you don't know exact dollar amounts for principle and interest, enter the entire payment as principle. But also, include a line for interest and give that line a 0 (zero) dollar amount. This will be adequate to let you issue the loan payment (print the check, etc.), but you'll need to come back to the transaction later and correct the principle and interest amounts.

Add a To Do list note (Lists|To Do Notes) as a reminder that you need to return to a particular transaction to make changes.

After you have the necessary information, return to the transaction and correct it. The transaction total won't need to be changed, just the portions assigned to principle and to interest.

The purpose of originally entering the interest portion of the payment as 0 (zero) is that you'll be more likely to notice it later, if you should forget to correct the transaction. If you print a transaction detail report or income tax report which includes the Interest & Finance Charges account, any transaction with a zero amount ought to raise a red flag for you. It should be much more noticeable than a transaction with an estimated interest amount—which would look just like all the other interest expense transactions.

Getting Information on Loan Payments and Unpaid Balances

Problem   *How can I get a list of the principle payments I've made on a loan? How can I find out the remaining unpaid principle balance?*

Solution   Use the Chart of Accounts window, and any of various reports to get loan payment details.

Finding the Unpaid Principle Balance

One of the benefits of setting up a separate account for each loan, is that the remaining unpaid principle balance of each loan will always be visible in the Chart of Accounts window. Here's part of a Chart of Accounts window, as an example:

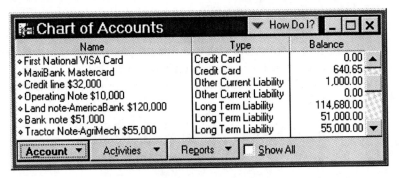

Also, you can open a Register window for any balance sheet account. Just double-click the account's name in the Chart of Accounts window. Or highlight the account name by clicking on it, then type Ctrl-R. The Register window for a loan account shows all of the loan's transactions—all borrowing and principle payments.

Getting a List of Loan Principle Payments

The easy way to get a list of all principle payments made on a loan, is to print a QuickReport for the loan account. There are a couple different ways to get a QuickReport. One is to open a Register window, then click on the QuickReport button at the bottom of the window. Or, you can highlight an account in the Chart of Accounts window, click on the Reports button at the bottom of the window, and then choose QuickReport from the pop-up menu. Here's a Quick-Report window for the Land Note-AmericaBank $120,000 account:

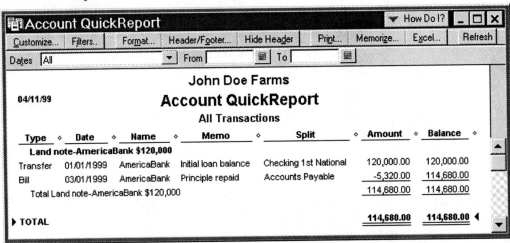

To view the transaction underlying any detail line of a , move the mouse cursor over the transaction until the cursor changes to a magnifying glass shape. When

320

it does, double-click the left mouse button. A window for the transaction will open, allowing you to view and/or edit it.

This page is
intentionally blank.

CHAPTER 12

Essential QuickBooks Reports

One of the most popular QuickBooks features is its reporting capability. QuickBooks reports are flexible, easy to use, and versatile. They allow "drilling down" to details of any transaction that's a part of any report total—which even makes them good for locating specific transactions to review or edit.

The QuickBooks User's Guide and Help system do a fine job of telling how to use QuickBooks' report features, so there's no reason to duplicate those instructions here. Instead, this chapter presents key points about just a few of QuickBooks' report features, plus information and ideas related to the most important reports for farm businesses.

Report Basics: The Profit and Loss Report

Problem   *What's a good way to get a report of income and expenses?*

Solution   Print a Profit and Loss report.

Discussion   As a summary of income, expenses, and net profit, the Profit and Loss report may be the most basic and universally used QuickBooks report. Whether you only keep simple income and expense records or you do a meticulous job of double-entry accounting, sooner or later you'll print a profit and loss report.

You probably already have an idea of what a profit and loss report is and what information it provides. So rather than focus on specifics of the Profit and Loss report itself, this section mostly uses that report to illustrate some common QuickBooks report features.

Can I prepare income taxes from a Profit and Loss report?

Though you could use the Profit and Loss report for income tax preparation, other QuickBooks reports are better suited to the job. See Preparing Income Tax Reports, on page 342.

Summary vs. Detail Reports

Most QuickBooks reports are variations of two basic styles: summary and detail.

Summary reports provide totals for accounts, classes, customers, vendors, or whatever is the subject of the report. Summary reports are the usual way accounting information is reported, and the balance sheet or profit and loss report you present to others (lenders, partners, farm management advisor, etc.) will normally be in summary form.

Detail reports provide totals too, but also list the individual transactions that contributed to them. Detail reports make it easy to find and correct accounting errors, such as income or expense posted to the wrong account, or an incorrect dollar amount. They are also good to print out as a permanent copy of transactions, for safekeeping.

For a printed copy of your transactions...

At the end of each month, and again at the end of the year, some QuickBooks users print a detailed report of transactions for the period and store the printout offsite (at some location away from the computer). The Transaction Detail by Account report (Reports|Transaction Detail Reports|By Account) is good for this purpose because it includes transactions for *all* accounts—assets, liabilities, equity, income, and expenses.

Opening a Profit and Loss Report

To open a window for the Profit and Loss report:

1. **Choose Reports|Profit and Loss|Standard (or one of the other Profit and Loss report variations).**

   The Profit and Loss report window will open. (Depending on how you have preferences set, the Customize Report dialog might actually open before the report window opens, to let you customize report settings first. Customizing reports is described later in this section.) Here's part of the Standard Profit and Loss report window:

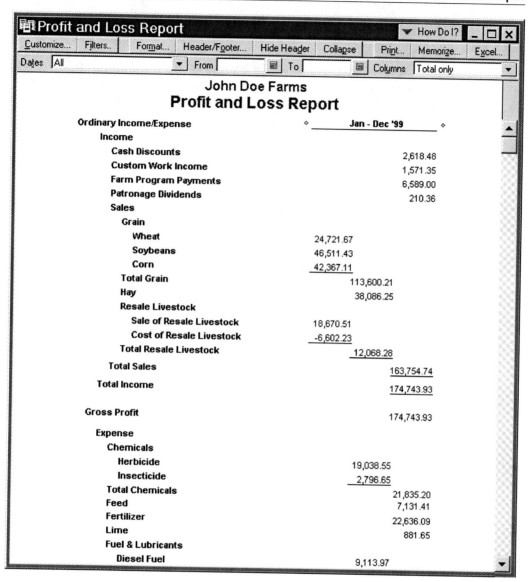

The buttons at the top of the window let you filter (limit) the types of transactions included, customize report settings, change report appearance and titles, print the report, and more. The selection boxes just below the buttons give you quick access to often-used report settings such as the date range of transaction included in the report.

Filtering Reports: Choosing What Gets Included

When a report is based on transactions other than the ones you want, you need to *filter* it, which means to specify or limit the set of transactions it includes. But also, filtering expands the usefulness of QuickBooks reports. You can use filters to limit a report to transactions involving a particular customer, vendor, account, class, month, week, day, dollar amount, and more—to help you do things like find all the transactions related to a billing mix-up, or figure out why the checking account won't reconcile.

You can quickly filter the range of transaction dates in a report by clicking the down arrow in the Dates field at the top of the report window and selecting a date. Or, you can set the date range precisely by selecting dates in the From and To fields. To filter the report in other ways you'll need to use the Report Filters dialog:

1. **Click on the Filters button at the top of the report window.**

   The Report Filters dialog will open.

2. **Set one or more report filters.**

   The Report Filters dialog varies a bit depending on which report you're using, but this example from a Profit and Loss report is typical:

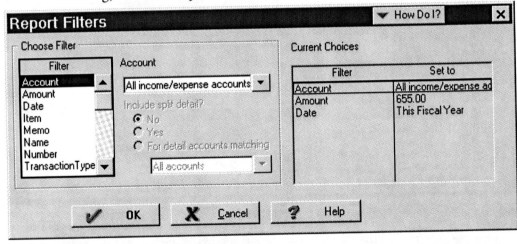

You can think of the Report Filters dialog as consisting of three parts: (1) the Filter box on the left, (2) the area in the middle where you specify the details for a filter, and (3) the Current Choices area on the right, which displays a quick summary the filters you have set. To set a filter, click on the desired field description in the Filter box. Then specify the filter's settings in the middle area, just to the right of the Filter box. An item will appear in the Current Choices list for each filter you set.

Three filters are set in the example shown above. The Account filter is automatically included by QuickBooks when preparing a Profit and Loss report. It limits the report to include only income and expense transactions. The Amount filter limits the report to transactions with a dollar amount of $655.00. (This is the kind of filter you might set to locate a specific transaction, when you know only one or two pieces of information about it, such as the date, payee, or amount.) The Date filter limits the report to transactions for the current year.

Customizing Reports: Controlling How Information is Displayed

You may change the amount of information a report displays and the arrangement of it by *customizing* the report. To customize a report:

1. Click on the Customize button at the top of the report window.

The Customize Report dialog will open.

2. Select the customization options you want.

Like the Report Filters dialog, the Customize Report dialog varies considerably depending on the report you are using. Here is an example of the Customize Report dialog for a Profit and Loss report:

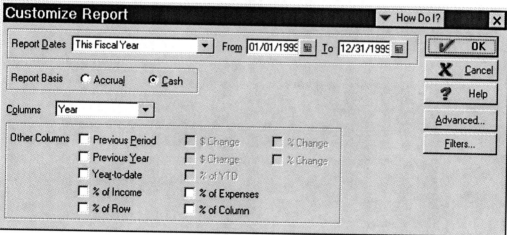

And here are comments about some of the fields:

Report Dates

The first thing you may notice about the Customize Report dialog is that it lets you select a transaction date range for the report, which you can also do in the main report window and from the Report Filters dialog. Because setting the date range is something you will use often, QuickBooks lets you do it from several different places.

Report Basis

This is the option that allows getting a cash basis report—such as a cash basis income tax report—from accrual records. If you are a cash basis record keeper you may want to always choose the Cash option.

Actually, switching between accrual and cash-basis reports only happens if you have some accrual transactions. From a QuickBooks point of view, accrual transactions are any that are posted to the Accounts Receivable or Accounts Payable accounts. You don't think you have any of these? Think again: if you use either the Bills form or the Invoices form, you have accrual transactions.

Switching between accrual and cash-basis reports

The sure way to get cash basis reports is to keep cash basis records. In other words, you only enter income when payment is received and expense when you actually pay for something.

But QuickBooks is by design an accrual record system, which means income is recorded whenever you enter an invoice (Activities|Create Invoices) and expense is recorded whenever you enter a bill (Activities|Enter Bills), regardless of whether any money has been paid. You may or may not use invoices, but most farm business users of QuickBooks use bills to manage accounts payable. If you do, your records include some accrual transactions. To assure getting cash basis reports from QuickBooks you'll need to use the Cash Report Basis option (in the Customize Report window, available by clicking a report's Customize button).

If you've selected the Cash Report Basis option, while preparing the report QuickBooks temporarily "reverses out" transactions posted to Accounts Payable and Accounts Receivable type accounts. If you've selected the Accrual report basis option, those transactions are included in the report as is.

Columns

This field lets you select a variety of different arrangements for columns of data in the report.

Other Columns

The Other Columns check boxes let you include optional columns, such as percentage calculations and prior-period or prior-year totals, to help you analyze the report's data and compare it with past business performance.

As this is written, no QuickBooks reports are capable of totaling **sales and purchase quantities**. Chapter 7, Quantity Information in Income & Expense Transactions, describes the options available for getting quantity information from your QuickBooks accounting records.

Changing a Report's Appearance

You can change almost everything included in a QuickBooks report, including the report title, headings, how numbers are displayed, and so on. To experiment with some of these "cosmetic" changes try the Format, Header/Footer, Hide Header, and Collapse buttons at the top of the report window.

Memorizing a Report

After spending lots of effort to get a report's settings just the way you want them it's a good idea to have QuickBooks memorize the report, which saves the report's settings in the Memorized Reports list. To memorize a report:

1. **Click on the Memorize button at the top of the report's window.**

   QuickBooks will display a Memorize Report dialog.

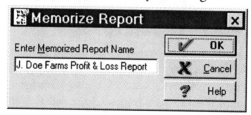

2. **Enter a name for the memorized report in the Memorize Report dialog.**

   Use any name that will help you recall the report's purpose when you see it in the Memorized Reports list.

3. **Click OK to close the Memorize Report dialog.**

   The report definition will be saved in the Memorize Reports list.

After memorizing a report, recreating it is simple. Just open the Memorized Reports List (Reports|Memorized Reports), then double-click the desired report name. Usually you'll only have to change a few report settings such as the transaction dates involved, to use the report for a different month, year, or other new situation.

QuickZoom: "Drilling Down" to Individual Transactions

One of the handiest QuickBooks report features for getting at detailed information is called QuickZoom. QuickZoom lets you "drill down" through layers of report totals to see the individual transactions which contributed to those totals. To use QuickZoom to get details about any amount or transaction in a report:

1. **In the report window, move the mouse cursor over the desired line or amount, until the cursor changes to a magnifying glass shape.**

    The change to a magnifying glass shape is QuickBooks' way of telling you that the report item the cursor is on can be viewed in detail.

2. **Double-click the left mouse button.**

    QuickBooks will open a new window containing details about the report item.

    If you double-clicked a transaction line in a detail report, the new window will be an edit window for the transaction you double-clicked. If you are working with a summary report, the new window will show a list of the transactions underlying the report item, and you may double-click any of the listed transactions to open an edit window for them.

    If you make changes in a transaction that affect the results of a report—such as changing a dollar amount—QuickBooks will automatically update the report window based on those changes.

Newer QuickBooks versions give you the option of **manually refreshing reports** when transactions are changed, by clicking on a Refresh button in the report window. Refreshing manually prevents having to wait while QuickBooks updates open report windows each time you change a transaction. For information search for "Refreshing reports and graphs" in the QuickBooks Help system.

Preparing a Balance Sheet Report

| **Problem** | *What are my options for preparing a farm balance sheet report?* |

| **Solution** | With consideration for the limitations of QuickBooks, you can either use one of QuickBooks' balance sheet reports or prepare the balance sheet with a spreadsheet program. |

| **Discussion** | Though QuickBooks is an innovative product in the way that it handles income and expense transactions, accounts payable and receivable (bills and invoices), and many other activities, you could say it takes a "bare bones" approach to |

accounting for assets, liabilities, and owner's equity. It has few features specifically designed to help with those parts of the accounting job necessary for maintaining a balance sheet. There are no special fixed (depreciable) asset management features, no provision for keeping a "dual" balance sheet (with assets listed both at book value and at market value), and no easy way to maintain the value of accounts representing farm inventories (grain, market livestock, supplies on hand, etc.).

But as the famous Muppet, Miss Piggy says, "Ya' gotta' go with whatcha' got!". There are ways to work around the limitations and use QuickBooks to prepare an acceptable farm balance sheet based either on book values or market values of assets.

The purpose of this section is to discuss actual preparation of the balance sheet report—not the accounting behind it. See chapter 10, Farm Asset Basics, for ideas and accounting techniques for maintaining market value and book value balance sheet in QuickBooks.

Market Value?...Book Value?...or Both?

QuickBooks only keeps one set of balance sheet figures—one balance for each account. How you do the accounting for asset values determines whether those account balances represent the book value or the market value of the farm's assets; and thus, whether the balance sheet you print from QuickBooks will be at book value or at market value.

If you need to prepare *both* a book value and a market value balance sheet, the easiest approach is to maintain book values in QuickBooks. The balance sheet reports you print from QuickBooks will then be at book value. A market value balance sheet can then be prepared by exporting a copy of a balance sheet report to a spreadsheet program. Once it's in spreadsheet form, you can substitute estimated market values the asset accounts and print a market value version of the balance sheet report, as described later in this section.

Preparing a Book Value Balance Sheet

Preparing a balance sheet report in QuickBooks is simple—only a couple menu clicks are necessary. What is not as simple, is doing all the necessary accounting beforehand. Here's a short summary of the steps:

1. **Enter fixed asset purchases and sales to properly maintain fixed asset accounts at book value.**

   You do this throughout the year of course. Those entries are discussed in chapter 10, Farm Asset Basics.

2. **Enter depreciation expense.**

   Most people do this once a year, after completing the year's accounting. Or, to prepare an accurate mid-year balance sheet, you may enter year-to-date depreciation expense just prior to preparing the balance sheet report. This was also discussed in chapter 10.

3. **Adjust the values of accounts representing farm inventories, to estimated book values.**

   From an accounting standpoint this step is something that should not happen. Almost by definition, book values should be maintained by the accounting system, not estimated. But QuickBooks' shortcomings in handling farm inventories makes adjustments to estimated book value the most practical approach. This topic was also covered in chapter 10.

4. **Open the balance sheet report.**

   Choose any of the report variations listed in the Reports|Balance Sheet menu.

   The illustration below shows part of a Summary Balance Sheet report window (Reports|Balance Sheet|Summary), with assets listed at book value.

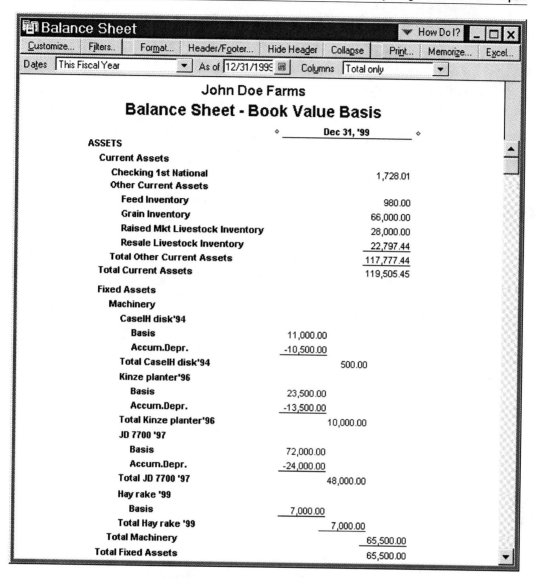

Notice that detailed accounts are used for fixed assets in this example. The purpose is not to suggest that everyone should necessarily use them, but to allow you to compare this balance sheet with the one shown on page 335. (That one is a market value balance sheet, and the detailed fixed asset accounts and subaccounts illustrate a technique for adjusting book values to market values.)

5. **Print the balance sheet report.**

Click on the Print button in the report window.

Preparing a Market Value Balance Sheet in QuickBooks

Preparing a market value balance sheet report in QuickBooks requires putting in some effort before the report can be printed. Here's a short summary of the steps:

1. **Adjust asset account balances to their estimated current market values.**

   This step is necessary for bringing farm asset values up to date, and was discussed in chapter 10, Farm Asset Basics.

2. **Open the balance sheet report.**

   Choose any of the report variations listed in the Reports|Balance Sheet menu.

   Here's a Summary Balance Sheet report window like the one shown earlier (Reports|Balance Sheet|Summary), but in this one the assets have been adjusted to estimated market values. (The market value adjustments have been made by posting a value to the Mkt Value Adjustment subaccount, as described in chapter 10.)

John Doe Farms
Balance Sheet - Market Value Basis

ASSETS	Dec 31, '99
Current Assets	
Checking 1st National	1,728.01
Other Current Assets	
Feed Inventory	900.00
Grain Inventory	58,000.00
Raised Mkt Livestock Inventory	18,000.00
Resale Livestock Inventory	22,797.44
Total Other Current Assets	99,697.44
Total Current Assets	101,425.45
Fixed Assets	
Machinery	
CaseIH disk'94	
Basis	11,000.00
Accum.Depr.	-10,500.00
Mkt Value Adjustment	5,500.00
Total CaseIH disk'94	6,000.00
Kinze planter'96	
Basis	23,500.00
Accum.Depr.	-13,500.00
Mkt Value Adjustment	2,000.00
Total Kinze planter'96	12,000.00
JD 7700 '97	
Basis	72,000.00
Accum.Depr.	-24,000.00
Mkt Value Adjustment	5,000.00
Total JD 7700 '97	53,000.00
Hay rake '99	
Basis	7,000.00
Total Hay rake '99	7,000.00
Total Machinery	78,000.00
Total Fixed Assets	78,000.00

3. Print the balance sheet report.

Click on the Print button in the report window.

Another way to prepare a market value balance sheet is to do it in a spreadsheet program. The basics of that process are described by the next topic, and a specific example is presented on page 338.

Exporting a Balance Sheet Report to a Spreadsheet Program

When a QuickBooks report does not contain all the necessary information or is not formatted as you want, one of your options is to send it to a spreadsheet program such as Microsoft Excel or Corel Quattro Pro, make additions or changes there, then print the spreadsheet version of the report. Sending a report to a spreadsheet is usually a two-step process: you (1) export a copy of the report from QuickBooks to a disk file, then (2) load or import the disk file into the spreadsheet program.

Beginning with the Pro release (but *not* the standard release) of Quick-Books 99 you can **send any QuickBooks report directly to a Microsoft Excel spreadsheet** without any manual steps! All Quick-Books Pro 99 report windows include an "Excel" button. Clicking it opens Microsoft Excel and transfers a copy of the report into it.

Here are the steps for manually exporting a balance sheet report from Quick-Books to a disk file. (These same steps work for exporting almost any Quick-Books report to a disk file.)

1. **In QuickBooks, open any balance sheet report.**

   To open QuickBooks' standard balance sheet report, choose Reports | Balance Sheet | Standard.

2. **Customize and filter the report to contain the information you want.**

   You may change any report settings you wish. Customizing and filtering reports were discussed earlier in this chapter.

3. **In the report window, click on the Print button.**

   QuickBooks will display the Print Reports dialog.

4. **Choose appropriate settings for printing the report to a file, in the Print Reports dialog.**

   The settings you choose in the Print Reports dialog control how QuickBooks will format (arrange) the report data it sends to a disk file. The instructions below mention two different file format options. If you don't yet know which one will work best in your particular spreadsheet program, just try one—you can decide which format is best by trial and error, seeing which is easiest to load into your spreadsheet program.

   This appearance of the Print Reports dialog differs considerably in older and newer QuickBooks versions, so here are instructions for both.

   • **If you use QuickBooks 4.0 or earlier:**

     This illustration shows only the upper part of the Print Report dialog, containing the fields that are discussed below:

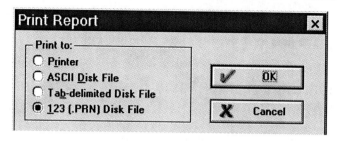

1. In the Print To: box, select either "Tab-delimited Disk File" or "123 (.PRN) Disk File".

2. Click OK.

   The Create Disk File window will open.

3. Enter a file name to use for the exported report.

4. Click OK.

   QuickBooks will send a copy of the report to the file.

* **If you use QuickBooks 5.0 or later:**

   This illustration shows only the upper part of the Print Reports dialog, containing the fields that are discussed below:

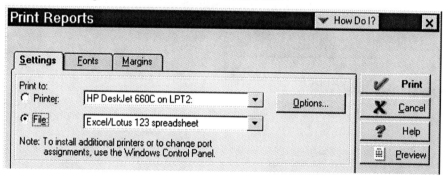

On the Settings tab of the Print Reports dialog:

1. Select the "File" Print To option, by clicking on it.

2. In the box to the right of "File", click the down arrow and select either "Tab-delimited file" or "Excel/Lotus 123 spreadsheet".

3. Click the Print button.

   The Create Disk File window will open.

4. Enter a file name to use for the exported report.

5. Click OK.

QuickBooks will send a copy of the report to the file.

The next topic describes how to import disk file into a spreadsheet program.

Preparing a Market Value Balance Sheet in a Spreadsheet Program

Many different spreadsheet programs are available. Microsoft Excel 97 was used for this example only because it is a popular spreadsheet program. The command details for using another spreadsheet program may differ, but the results would be similar.

Importing a Balance Sheet File into Microsoft Excel

Here are the steps for opening a balance sheet report file in Microsoft Excel 97, assuming the file was exported to disk from QuickBooks as a .PRN file (using either the "123 (.PRN) Disk File" or "Excel/Lotus 123 spreadsheet" print options, as described on the previous page):

1. **Open the Microsoft Excel program.**

2. **Choose File|Open in Microsoft Excel.**

   A file open dialog will be displayed.

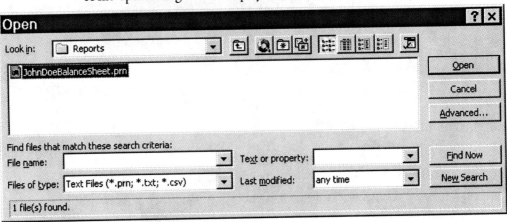

3. **Select and open the .PRN file.**

   - In the "Files of type" field at the bottom of the file pen dialog, select "Text files (*.prn; *.txt; *.csv)".

   - If the .PRN file is located in some folder other than where Microsoft Excel first looks, switch to that folder by using the Look In box at the top of the file open dialog.

- When you find the desired file, highlight it by clicking on it. Then click the Open button.

Because a .PRN file is not a standard Excel file, Excel will display the first dialog panel of the Text Import Wizard to guide you through the process of loading the file.

Original data type

Choose the file type that best describes your data:

- ⊙ Delimited   - Characters such as commas or tabs separate each field.
- ○ Fixed width   - Fields are aligned in columns with spaces between each field.

4. In the first wizard panel choose Delimited.

Leave the other dialog options as they are. Click on the Next button (not shown) to move to the second panel.

Delimiters

- ☐ Tab   ☐ Semicolon   ☑ Comma
- ☐ Space   ☐ Other:

☐ Treat consecutive delimiters as one

Text Qualifier: "

5. In the second wizard panel select Comma as the delimiter by clicking on the Comma check box.

Leave the other options as they are. Click on the Next button (not shown) to move to the third panel.

This screen lets you select each column and set the Data Format.

'General' converts numeric values to numbers, date values to dates, and all remaining values to text.

Column data format

- ⊙ General
- ○ Text
- ○ Date: MDY
- ○ Do not import column (Skip)

6. In the third wizard panel leave the Column data format set at General.

7. Still in the third wizard panel click on the Finish button (not shown) to load the file.

The file might look something like this after loading into Excel:

	A	B	C
1		Dec 31, 99	
2	ASSETS		
3	Current Assets		
4	Checking/Savings		
5	Checking	1728.01	
6	Total Chec	1728.01	
7	Other Current Assets		
8	Feed Inven	980	
9	Grain Inver	66000	
10	Raised Mk	28000	
11	Resale Liv	22797.44	
12	Total Othe	117777.4	
13	Total Curre	119505.5	

Obviously, the imported file needs reformatting to make it more readable.

Reformatting the Imported File and Updating Market Values

Here are suggested steps for reformatting the report file after importing into Microsoft Excel, and for updating asset values:

1. **Widen the data columns.**

   Move the mouse until the cursor is over the divider between the A and B column labels. Click and hold down the left mouse button, and drag the mouse to the right to widen column A—enough so you can read all of the descriptive text in column A—then release the mouse button. Do the same for the divider between columns B and C to widen column B.

2. **Change the number format of column B to Currency.**

   Click once on the B column label to highlight the entire column. Then choose Format|Cells. Excel will open a Format Cells dialog. Click on the Number tab, then in the Category box select Currency. Click OK to close the Format Cells dialog.

3. **Change values in column B, where appropriate, to reflect current asset market values.**

4. **Substitute spreadsheet formulas, where appropriate, so the spreadsheet will calculate and display correct subtotals and totals.**

You'll need to add spreadsheet formulas to total the spreadsheet cells which contain changed information, otherwise the balance sheet totals won't reflect the updated asset values and other changes you make.

If you aren't familiar with spreadsheet formulas, look up the SUM or @SUM formulas in your spreadsheet program's Help system.

5. **Make other changes and additions to the spreadsheet as desired.**

You may want to insert spreadsheet rows and add a report title, report date, or other report features. You may also delete rows that aren't needed.

It may be necessary to insert some rows for assets not included in the Quick-Books report. For example, fixed assets which are "depreciated out" (have a book value of zero) may not have been included in the QuickBooks balance sheet report, but may still have considerable market value and should be included in the market value balance sheet.

Here's an example of a finished spreadsheet version of the balance sheet, with grid lines turned off for improved appearance.

	A B	C	D	E
1		John Doe Farms		
2		**Balance Sheet - Market Value Basis**		
3	ASSETS			*Dec 31, '99*
4	Current Assets			
5	Checking/Savings			
6	Checking 1st National		1,728.01	
7	Total Checking/Savings			1,728.01
8	Other Current Assets			
9	Feed Inventory		900.00	
10	Grain Inventory		58,000.00	
11	Raised Mkt Livestock Inventory		18,000.00	
12	Resale Livestock Inventory		22,797.44	
13	Total Other Current Assets			99,697.44
14	**Total Current Assets**			**101,425.45**
15	**Fixed Assets**			
16	Machinery			
17	CaseIH disk94		6,000.00	
18	Kinze planter96		12,000.00	
19	JD 7700 97		53,000.00	
20	Hay rake 99		7,000.00	
21	Total Machinery			78,000.00
22	**Total Fixed Assets**			**78,000.00**
23	Other Assets			
24	Pat. Div. Asset Withheld		66.00	
25	Total Other Assets			66.00
26	**TOTAL ASSETS**			**179,491.45**

Preparing Income Tax Reports

Problem *What QuickBooks reports should I print to take to my tax preparer?*

Solution An Income Tax report and a listing of fixed asset purchases and sales should provide most of the information necessary for preparing your income tax return.

Discussion Some farm record systems are based on a chart of accounts that appears to be a clone of the categories on Form 1040, Schedule F—the Farm Income and Expenses tax form. Though there's nothing terribly wrong with that, the income and expense categories provided by the IRS are seldom as useful for gathering management information as they could be, primarily because they are not tailored to your farm business.

QuickBooks gives you the best of both worlds. You can set up the chart of accounts with income and expense categories that are the most useful to you, for management purposes, and you can also prepare income tax reports with income and expenses grouped according to the Schedule F categories.

At least two reports are usually necessary for tax preparation. One is the Income Tax report—the special QuickBooks report which groups your accounts according to IRS categories. The other report is a list of fixed asset purchases, sales, and disposals. Additional reports may also be required, depending on how you keep records.

The examples in this section focus on the tax reports needed for a sole proprietorship. Additional reports may be necessary if the farm business operates as a partnership or corporation. Ask your tax preparer about the reports he or she will need for your farm business.

Turning On the Tax Reporting Features (Picking a Tax Form)

None of QuickBooks' tax reporting features are available until you've told QuickBooks which IRS tax form your business files. Normally this is done when you set up a QuickBooks company, as part of the EasyStep Interview, but you can specify the tax form at any time, as follows:

1. **Choose File|Company Info.**

   The Company Information window will open.

2. Select a tax form in the Income Tax Form Used list box:

Income Tax Form Used	Form 1040 (Sole Proprietor) ▼

Which tax form for partnerships?

On Federal tax returns, farm income and expenses reported for a partnership are reported on a Schedule F form—the same form used by sole proprietorships. So instead of selecting the Partnership form 1065, selecting Form 1040 in QuickBooks may be a better choice if your farm business operates as a partnership. However, if the partnership involves various business enterprises in addition to farming, selecting the Partnership form 1065 may be best. It all boils down to deciding which approach will save you the most time when preparing reports to take to your tax preparer.

Assigning Schedule F Tax Lines to Income and Expense Accounts

After selecting a tax form as described above, QuickBooks will allow assigning a Schedule F Tax Line to new and existing accounts in your chart of accounts. Here's how to assign a tax line to an existing account:

1. **Open the Chart of Accounts window.**

   Choosing Lists | Chart of Accounts is one way to do this.

2. **Highlight the account you want to work with, by clicking on it.**

3. **Open the Edit Account window for the highlighted account**

   Either type Ctrl-E, or choose Edit | Edit Account, or click on the Account button in the lower part of the Chart of Accounts window and then choose Edit from the pop-up menu.

4. **In the Tax Line field, select an appropriate tax line for the account.**

   The Tax Line field of the Edit Account window looks like this:

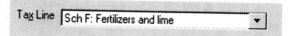

Tax Line	Sch F: Fertilizers and lime ▼

Note: *if your Edit Account window doesn't have a Tax Line field, you need to turn on QuickBooks' income tax reporting features by picking a tax form, as described earlier in this section.*

5. **Click OK to close the Edit Account window.**

343

6. **Repeat steps 2 through 5 to assign a tax line to other accounts.**

    Not all accounts should be assigned a tax line. In a typical sole proprietorship, tax lines are assigned only to income and expense accounts.

Verifying the Tax Line Assignments

QuickBooks versions 5.0 and later have an Income Tax Preparation report that you can use for verifying tax line assignments. The report lists each account and the tax line that's assigned to it. Here's how to get the report:

1. **Open the Chart of Accounts window.**

    Choosing Lists | Chart of Accounts is one way to do this.

2. **Click on the Reports button at the bottom of the Chart of Accounts window.**

3. **In the pop-up menu, choose Income Tax Preparation List.**

    QuickBooks will open the report window, which looks something like this:

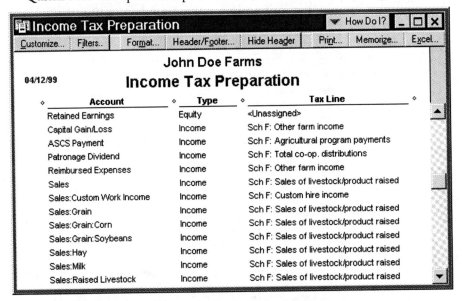

If your version of QuickBooks doesn't have an Income Tax Preparation report, print a chart of accounts listing instead. The standard report for the Chart of Accounts window includes the account's name and type, and also the assigned income tax line. To print the listing either (1) click the Accounts button in the Chart of Accounts window, then choose the Print List option from pop-up menu, or (2) click on the Reports button in the Chart of Accounts window, then choose Accounts Listing from the pop-up menu.

Preparing an Income Tax Report

With tax lines assigned to individual accounts, printing an income tax report is simple:

1. **Open the desired income tax report window.**

   QuickBooks has both a summary and a detail version of the Income Tax report, so choose either Reports|Other Reports|Income Tax Summary or Reports|Other Reports|Income Tax Detail.

2. **Customize and filter the report to include the desired transactions.**

   Be sure the transaction date range matches your tax year. Also, be sure the Report Basis option (Cash or Accrual) matches your tax filing method.

3. **Check for income or expense which is not assigned to any tax line.**

   The first step you should take after printing a tax report is to check for "unassigned" income or expense. If any of your transactions use accounts which have not been assigned a tax line, they will be grouped together near the bottom of the Income Tax report under the heading "Tax Line Unassigned".

   If you find "Tax Line Unassigned" amounts in the Income Tax Report, you need to determine which accounts are involved and assign tax lines to them. One way to find out which accounts are involved is to use the QuickZoom feature (described earlier in this chapter) to "drill down" to the individual transactions contributing to it, to learn which accounts are involved. (If you open an Income Tax Detail report window, the individual transactions and their account information will all be listed there on the report.)

   Here's a fragment of the Income Tax Detail report which shows that one account—Finance Charges—apparently has not been assigned a tax line:

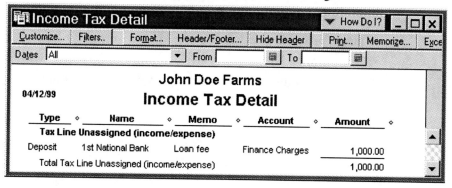

With this information, you could go to the Chart of Accounts window, find the Finance Charges account, and assign a tax line to it.

The Tax Line Unassigned section will no longer be present in the Income Tax report, once you have assigned a tax line to every income and expense account.

4. Print the Income Tax report to have a copy for your tax preparer.

Click on the Print button in the Income Tax report window.

Which Income Tax report should I print?

Printing both the detail and summary report versions works well.

Errors are easiest to find in the detail version, because it shows the transactions which contributed to each tax line's total. So you may want to print a copy of the Income Tax Detail report first and spend some time pouring over it. (This may be your last opportunity to catch errors in your records before they become part of your income tax return.)

The summary version of the Income Tax report may be best for taking to your tax preparer. He or she will be mostly interested in income and expense totals, not the individual transactions behind them.

Finally, you may want to keep a printed copy of the Income Tax Detail report to have as a record of the transactions on which your income tax filing is based.

Getting a Report of Fixed Asset Purchases/Sales/Disposals

Another report you will need is a list of fixed asset purchases, sales, and disposals for the year, so the tax preparer can add or remove items from the farm's depreciation schedule and can perform the necessary tax calculations.

With all fixed asset transactions posted to accounts of the Fixed Asset account type, getting the report is easy. Here's how:

1. Open the QuickBooks report you want to use for printing the list of fixed asset transactions.

The Transaction Detail by Account report is a good choice. Open it by choosing Reports | Transaction Detail Reports | By Account.

2. Filter the report to include only transactions which (1) involve Fixed Asset type accounts, and (2) are dated during the appropriate tax year.

Click on the Filters button in the report window to open the Report Filters dialog.

In the Filter box click on Account, then in the field to the right of the Filter box click the down arrow and select "All fixed assets".

In the Filter box click on Date, then in the field to the right of the Filter box click the down arrow and select the appropriate date description—usually Last Tax Year or Last Fiscal Year is the appropriate choice.

The Report Filters dialog should look something like this when you're done:

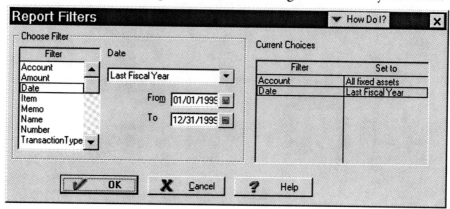

3. **Click OK to close the Report Filters dialog.**

The report should now list only the tax year's fixed asset transactions. Here's an example:

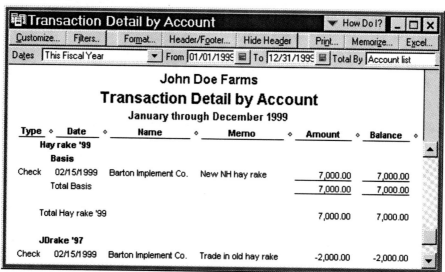

Note: this report will also include depreciation expense transactions if you have entered them. You may want to cross out those lines on the report copy you give to your tax preparer, to prevent confusion.

4. Check the report for errors.

5. Print the report.

Click on the Print button in the report window.

It's a good idea to print two copies of the report: one for your tax preparer, and one to keep with the farm business records.

CHAPTER 13

QuickBooks "Housekeeping"

Using almost any computer software involves some amount of "housekeeping"—taking care of little maintenance chores that keep the program running smoothly or protect data from loss. This chapter deals with topics related to QuickBooks program and file maintenance.

QuickBooks Data Files

Problem

What are .QBW files? .QBB files? .QBI files?

Solution

All are different types of QuickBooks data files.

Discussion

All data for a QuickBooks company is maintained in a *single* disk file, not multiple files as it is in some other programs. This makes managing QuickBooks data files easy. You can move them to different folders (directories), make backup copies, or delete data files you no longer use, all without worrying whether you've gotten "all" of the necessary files. To do these things though, you need to understand the purpose of the various types of QuickBooks data files.

.QBW Files

.QBW files are the main type of data files you will use in QuickBooks. They are often called "company files", because each .QBW file contains all of the records and setup information for a single company. To work with a different company's records in QuickBooks, you open a different company file. There is no limit to the number of company files you may work with in QuickBooks (one at a time, of course).

Here are other points to note about .QBW files:

+ **.QBW files are not location specific.**

  Though QuickBooks normally stores them in the same folder (directory) where the main QuickBooks program file is located, there's no reason to prevent moving them to some other location. Moving a .QBW file won't hurt anything so long as you understand how to find and open the file in its new location.

+ **You may rename any .QBW file without causing problems.**

  QuickBooks will still be able to open and use the file as before. However, only change the first part of the file name (the part before the period "."). Leave the file name extension (the ".QBW" part) intact so Microsoft Windows and QuickBooks can identify the file's type.

+ **You can copy .QBW files.**

  Another thing you can do with .QBW files is make duplicate copies of them in other folders (directories) on your hard disk, or even in the same folder if you give the copy a different file name. For instance, you might copy a .QBW file into a different folder or onto a diskette to save as an "archival" copy of the year's accounting records.

.QBB Files

.QBB files are QuickBooks's own form of backup files. When you choose File | Back Up in QuickBooks, this is the type of file QuickBooks creates.

A .QBB file is a "compressed" copy of a .QBW file, meaning the .QBB file contains the same information but in a form that takes much less space. Though the amount of compression varies, .QBB files are frequently one-fourth as large as the .QBW files they were created from. Their small size makes .QBB files a good choice for storing QuickBooks data long term. You may be able to store several years' data or several different company's files on a single diskette if they are stored as .QBB files.

Other points to note:

+ **.QBB files are not location specific.**

  They can be created, moved, or copied anywhere.

+ **You can rename .QBB files without causing problems.**

  Some users keep an archival copy of old transaction details by making a backup of their company (.QBW) file and renaming the backup file to include the year, such as FARM1998.QBB.

  See page 351 for tips on backing up your QuickBooks data.

.QBI Files

A .QBI file is a data file _for QuickBooks' internal use_. QuickBooks creates a .QBI file when you open a company (.QBW) file and deletes the .QBI file when you close the company.

The presence of a .QBI file on your hard disk means one of two things: either QuickBooks is running, or the QuickBooks program has "crashed" without getting an opportunity to delete the .QBI file.

Under normal circumstances you should _never_ touch a .QBI file—don't move it, don't rename it, don't delete it. QuickBooks will delete the file automatically the next time it opens the company. .QBI files may be deleted in special situations, such as to get rid of an old .QBI file which remains after renaming a company file, but _only delete .QBI files if you're sure QuickBooks is_ _not_ _running!_

Backing Up Your Records

Problem
How should I backup my QuickBooks records? How often? Where should I keep the backup copies?

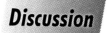
Solution
Backing up your records (making copies of them for safekeeping) is easy, because QuickBooks has a backup command. Read on to learn more.

Discussion
Making backup copies of important information is an essential part of using a computer. This is especially true for accounting records because of (1) the amount of time invested in entering records, and (2) the fact that they are so important—almost irreplaceable—for business and tax reasons. Luckily QuickBooks can create its own backup copies of your records, so make backup copies doesn't take much time or effort.

But besides knowing _how_ to make backups, you need to think about _when_ to make them and _where_ to keep them. You need an organized plan for making and storing backups to assure you'll have a good, useable copy of your records in case a software problem, computer crash, or natural disaster (flood, fire, etc.) renders the working copy of your company file useless.

What Media Should I Use for Backups? ...Diskettes? ...Tape?

Most of the time you should create backups on some kind of removable media, such as diskettes, tape cartridges, or Zip disks. Another relatively new possibility is Web-based backups. For a monthly service fee you can have your backups stored on a remote computer—somewhere at the other end of an Internet connection—in a secure, encrypted format (scrambled so that no one else should be able to read or use your data).

The important point is to make backups on something that can be physically moved to some location *other than* where your computer is. With backup copies stored off site (away from the computer), your valuable data should be safe from fire, flood, wind storm, and vandalism.

Making backup copies on your hard disk is also possible...but doesn't go very far toward protecting your data. If the hard disk should quit working, any backup copies stored on it would be no more accessible than the "live" copy.

Should I ever create a QuickBooks backup on my hard disk?

Yes, it makes sense in some situations. One is to keep an easily accessible copy of prior-year records available without occupying much hard disk space. (QuickBooks backup files are very compact.) Just be sure the hard disk copy is not the only backup of your records!

Another is when you want to make a "throw-away" backup copy. Before certain activities QuickBooks forces you to backup your data, even if you don't want to. If you already have enough backup copies of your records, it doesn't hurt to send the QuickBooks-required backup to your hard drive. Backing up data to a hard drive is often faster than backing it up to other media such as a diskette. Also, the backup file will be easily accessible for deletion, assuming you don't want to keep it.

How Do I Backup My Accounting Records?

To make a backup copy of the currently-open QuickBooks company file:

1. **Choose File|Back Up.**

   QuickBooks will display a file open dialog with the title "Back Up Company to...".

2. **Select a location (diskette drive, tape drive, Zip drive, etc.) and file name for the backup.**

   Always use the .QBB file name extension!

3. If you are using removable media, such as a diskette, place one in the specified drive.

4. Click the Save button (OK button in some QuickBooks versions).

QuickBooks will create the backup file using the drive and file name you specified.

If you are backing up to diskettes...

Always begin with plenty of blank formatted diskettes—QuickBooks won't format diskettes for you, and it won't overwrite existing files on a diskette. If your company file grows large enough (to around seven megabytes in size), QuickBooks may ask for additional diskettes to complete the backup, so have an extra one ready just in case.

Labeling backup diskettes

It's important that you label your backup copies to keep them in order.

If QuickBooks requests additional diskettes to complete a backup, write a sequence number of each diskette. This will make it easier to begin with the right diskette if you ever need to restore your data from a backup.

Also, each set of backup diskettes should be uniquely identified, even if the set contains just one diskette. That will prevent a possible mix up of diskettes from different backup sets. Diskettes in one set could be labeled A-1, A-2, A-3, etc. Diskettes in a second set could be labeled B-1, B-2, B-3, etc. And so on.

When (How Often) Should I Backup My Records?

How often you make backups is a matter of personal choice. But deciding on a backup frequency is not hard if you think of it in terms of how much your time is worth, and the amount of time that might be required to replace lost records.

For example, suppose you have made just a few QuickBooks entries since your last backup, and you have printed documentation for each of them—check stubs, invoices, etc. Because you've spent little time entering transactions, and because each of them would be easy to duplicate if they had to be re-entered, making a backup isn't too important.

Now compare that to a time when you've spent hours making entries, or you've entered some transactions that would require a lot of effort to duplicate if you had to re-enter them. In this situation, spending a few minutes to make a backup of your records would obviously be worthwhile—cheap insurance against the loss of valuable records and the time you have invested.

OK, so we can all agree that it's important to make backups often, assuming that *how* often relates to the amount of time you spend entering records. But there's another problem. Without a regular schedule or plan for making backups it's too easy to either forget to make them, or put the job off until another day "when I'm not as busy".

What's a Good Backup Plan to Follow?

A good backup plan consists of these parts: (1) backup frequency, (2) a regular schedule, (3) a sufficient number of backup sets, (4) rotating the backup sets, and (5) storage locations for the backup sets.

- The **backup frequency** that's right for you should be based on the amount of time you spend entering data. If you average an hour a day making entries in QuickBooks, then you probably should make backups every day or two. If you spend no more than a couple hours per week, then a weekly backup may be enough.

- Once you've decided on the backup frequency (daily, weekly, bi-weekly, etc.), develop a **regular schedule** for making backups. For example, if you've decided on a weekly schedule, get into the habit of always making backups on a certain day of the week, such as Friday. And if you aren't able to make your regular backup on the chosen day, get back on schedule by making a backup as soon as possible.

- A critical part of every good backup plan is to have a **sufficient number of backup sets**. Don't make the mistake of having just one backup copy of your records. If you need to restore from it, there is no guarantee it will be readable. Always keep at least three, and preferably five or more backups sets. And be sure to label them to keep them separate.

- Be sure to follow a regular plan for **rotating the backup sets**. If you have five backup sets labeled A, B, C, D, and E, try to use them in that order. And after using set E, start over with set A.

- Finally, the **storage location** for the backup sets is important. "All of them in the left-hand desk drawer" is *not* a good plan. As mentioned earlier, it is important to have at least some of the backup copies off site, at some location other than where the computer is. That may be at a relative's house, or in a safety deposit box, or in a metal box out in the machine shed or shop building. Using any physically separate location is the key to protecting against losing data to fire, theft, and other hazards. Because it may be more difficult to include remotely-stored backup copies in the regular rotation of backup sets, you my update them on a less-frequent schedule, such as once per month.

When Something Seems Wrong: Verifying/Rebuilding/ Restoring Data Files

Problem

I have a feeling that something is wrong with either the QuickBooks program or my data file. Some transactions appear to be missing [...or any number of other possible symptoms]. What should I do?

Solution

Use QuickBooks' features for checking and rebuilding your data file. If that fails, try restoring the data from a Quick-Books backup.

Discussion

Any number of problems are possible reasons for a mal-function in QuickBooks or in your QuickBooks data. Your computer or hard disk may be failing, or QuickBooks could have a software conflict with another program, or the oper-ating system (i.e., Microsoft Windows) may be working incorrectly, or its file system may be damaged, and the list goes on. It may not be comforting to hear, but problems with QuickBooks data are often due to random occurrences— things that won't necessarily happen again.

If you ever get the feeling that some of your QuickBooks data is "not quite right", the best thing to do is let QuickBooks verify the integrity of the data file. If problems are found, you can let QuickBooks make an attempt at rebuilding the data file on its own. And if that fails, you can restore the data from a backup copy. (You *have* been making backups, haven't you?)

How to Verify a QuickBooks Data File

When QuickBooks verifies a data file it merely checks the file for problems. If no problems are found, you may assume the symptoms you've noticed are due to something other than a data file problem. To verify a QuickBooks data file:

1. **Open the QuickBooks data (company) file you wish to verify.**

2. **Close any windows open in QuickBooks by choosing Windows|Close All.**

   Verify is found on the File|Utilities menu, which is only accessible if all open windows in QuickBooks have been closed.

3. **Choose File|Utilities|Verify Data.**

QuickBooks will check the file for errors—the process could take a while if the file is large. Afterwards, QuickBooks will display a message telling you whether problems were found. If so, try rebuilding the data file to see if that can fix the problem.

How to Rebuild a QuickBooks Data File

If verifying the data file found problems, try rebuilding the file:

1. **Open the QuickBooks data (company) file you wish to rebuild, if it is not already open.**

2. **Close any windows open in QuickBooks by choosing Windows|Close All.**

   Rebuild is found on the File|Utilities menu, which is only accessible if all open windows in QuickBooks have been closed.

3. **Choose File|Utilities|Rebuild Data.**

4. **Make a backup of the data file when prompted by QuickBooks.**

   QuickBooks requires making a backup of the company file as a safety measure. That way, if the rebuild operation fails you may at least be able to salvage part of your data from the backup.

   After completing the backup QuickBooks will attempt to rebuild the file. If the rebuild is successful, you may want to try the verify operation once more to be sure the file has no errors. If the rebuild was not successful, you may want to restore the data file from a backup.

How to Restore QuickBooks Data from a Backup Copy

Here are the basic steps for restoring data from a QuickBooks backup (.QBB) file.

1. **Choose File|Restore.**

   QuickBooks will display a file open dialog titled "Restore From".

2. **Select the location and name of the .QBB backup file to restore.**

   The file may be located on a diskette, for example.

3. **Click the Open button.**

   QuickBooks will display a file save dialog titled "Restore To", with the original company file's name pre-entered in the File Name box.

4. **Type a *different file name* in the File name box, but keep the .QBW file name extension.**

Never restore from a backup to an existing data file—even if that file is damaged! Parts of the file may be salvageable, but restoring to it will completely erase the file. If the restored data is in worse condition, you will wish you still had the option of opening the original damaged file.

5. **Click the Save button.**

QuickBooks will restore the data from the backup and, when finished, will display a message letting you know if the restore was successful.

Consult the QuickBooks Help system for more detailed information about restoring QuickBooks data from a backup file.

The last resort: abandon ship...but salvage as much as you can!

There's a sad, sinking feeling that comes from realizing that all your attempts to rebuild and restore your QuickBooks data have failed, and you are either going to have to start over with a new QuickBooks company file or paint a bulls-eye on your computer and use it for target practice. But there's one more thing to try.

Try opening the damaged data file in QuickBooks. If parts of the file are still accessible—such as the Customer list, Vendor list, Chart of Accounts, Class list, and other lists—you can at least salvage those parts. Try exporting all of the lists from the damaged file, creating a new company file, and then importing those lists as a starting point for building the new company file. (See the QuickBooks Help system for information on exporting and importing lists.)

Keeping Transaction Details: Archiving and Condensing the Company File

Problem

I want to keep transaction details available in QuickBooks as long as possible—so I can look back at transactions that are a couple years old if I want. But will my company file grow too large if I keep that much "live" data in it? And how can I remove old transactions from the company file?

Solution

Use QuickBooks' Condense feature to remove old data from the company file. But first, make an archival copy of the file to keep in case you want to look at transaction details from a prior year.

Discussion

QuickBooks has tremendous capacity for keeping old transaction data "live"—available and directly accessible—for years. Depending on the number of transactions you enter, it might be possible to keep three to five years or more of transaction details in the farm's company file. So capacity is not a problem...but speed is. The more transaction detail retained in a company file, the slower QuickBooks works. Report generation and lookup operations (such as the Find command) can become excruciatingly slow even on a fast computer. Also, as the company file grows larger, making backups takes more time and more diskettes.

The solution is to use QuickBooks' Condense Data feature to remove old transaction detail. QuickBooks lets you choose a cut-off date for the condense, then reduces all transactions prior to that date to a single General Journal entry for each month—a whole year's detail gets reduced to twelve General Journal entries. Condensing maintains proper account balances but gets rid of the detail that contributed to them.

There's no need to get rid of old transaction detail entirely, though. Before condensing a company file you should make a copy of it. Then you can open this archival copy in QuickBooks whenever you want to look at prior-year transactions.

Realistic expectations: How much "live" detail can I keep?

With typical computer hardware and a moderate number of transaction entries, most farm businesses should be able to keep at least two years of transaction detail without a major slowdown in QuickBooks operation. With a fast computer and a light load of transactions, keeping up to five years of detail may be possible.

Maintaining Archival Copies of QuickBooks Company Files

To make an archival copy of a company file, simply make a copy of the file using a new file name—keeping the .QBW file name extension, of course, so the archival copy can also be opened by QuickBooks. For example, if the farm's company file is named FARM.QBW, you might copy it using a name like FARMXXXX.QBW (where "XXXX" are the digits of a year, such as "1997").

Then if you condensed FARM.QBW to remove detail for the year 1997, you'd still have that detail available in the archival copy, FARM1997.QBW.

Unless your farm business has an unusually large number of transactions, it's best if you can keep *at least* the prior year's transaction details "live" in the company file. To accomplish this, at the beginning of each year you could archive and condense transaction data more than one year old.

For example, at the beginning of 2000 you could copy FARM.QBW to a file named FARM1997.QBW, then remove 1997 detail by condensing through 12/31/1997. Throughout 2000 you would have 1998's transaction detail records available in the FARM.QBW company file. If you needed to look up something in 1997, you would open the FARM1997.QBW file.

Unless you are short of hard disk space you may want to keep the archival copies on your hard disk, to make them readily available. So the archival copies aren't opened accidentally, it's a good idea to move them into a different folder on your hard disk—separate from the one where the farm's main company file is located.

Tips for Viewing/Using Archived Files

An archival copy is just like any other .QBW file—you can open it in QuickBooks, look at the transactions, print reports, etc. Actually, that's the problem. It is easy to forget that you are working with an archival copy and not the current company file. Here are steps that can help:

- Remember that any changes you make in the archival file will not appear in the current company file. Adding new transactions there, for instance, would be a waste of time.

- When you are done using an archival file, it is always a good idea to either close the file (File|Close Company) or re-open the *current* company file. Either way, you'll be insuring that the archival file won't be automatically loaded the next time you start QuickBooks—which could lead to making entries in the wrong file.

How to Condense QuickBooks Data

To condense old detail in a company file:

1. Open the company file you wish to condense.

2. Close any windows open in QuickBooks by choosing Windows|Close All.

Condense Data is found on the File|Utilities menu, which is only accessible if all open windows in QuickBooks have been closed.

3. **Choose File|Utilities|Condense Data.**

QuickBooks will display the Condense Data dialog.

4. **Select a cut-off date and the types of unused items you will allow QuickBooks to remove from the company file.**

Enter the cut-off date for condensing old transactions in the "Summarize transactions on or before" box. Select items to remove by checking the appropriate boxes. (It is safe to allow removal of all unused items, but you may not want to in all cases.)

Consult the QuickBooks Help system for more information about completing the Condense Data dialog.

5. **Click OK to close the Condense Data dialog.**

QuickBooks may display other messages or warnings, but eventually will require making a backup of the company file as a safety measure. That way, if the condense operation fails you will have a backup copy from which to restore the company file.

After completing the backup QuickBooks will condense transaction detail older then the cut-off date you entered. This can take quite a while if the company file contains many transactions. When it is finished, QuickBooks will display a message telling you whether the condense operation was successful.

Appendix A: Other Resources & Information

QuickBooks-Related Internet Sites

Intuit	www.intuit.com Current QuickBooks product and version information, FAQs (frequently asked questions), information about support program, ordering of checks and other paper forms, and more. (Intuit is the developer of QuickBooks.)
Flagship Technologies	www.agriculture.com/markets/flagship Information about using Quicken and QuickBooks for farm accounting, links to other Web sites with farm accounting information related to Quicken and QuickBooks, and information about the ManagePLUS add-on.
HelpTalk Forum	www.helptalk.com/qbpro/index.html A question-and-answer forum frequented by QuickBooks experts—an excellent place to get your *specific* QuickBooks questions answered, often in less than 24 hours, at no cost.
QuickBooks User-to-User Forum	www.intuit.com/quickbooks/forums/index.html Another site where you may post questions and receive answers, but less active than the HelpTalk Forum.

QuickBooks-Related Software

Here are references to selected software that enhances or adds to QuickBooks.

ManagePLUS

ManagePLUS is a software product which adds farm accounting features to QuickBooks. Developed by Flagship Technologies, Inc., it enhances capabilities for entering and reporting quantity information in QuickBooks transactions, and provides enhanced features for getting enterprise cost and revenue information, based on QuickBooks classes. A version of ManagePLUS is also available which works with Quicken.

For more information about ManagePLUS visit Flagship Technologies' Web site at www.agriculture.com/markets/flagship or call 800-545-5380.

FINPACK

FINPACK is a whole-farm business analysis software package developed and maintained by the Center for Farm Financial Management at the University of Minnesota. At this writing, FINPACK does not offer a direct interface for importing QuickBooks records. However, it is a popular farm financial analysis package and is used by many farmers and agricultural lenders.

For information about FINPACK visit the Center's Web site at www.cffm.umn.edu/cffm or call 800-234-1111.

Tax preparation software packages

TurboTax is software developed by Intuit, the maker of QuickBooks and Quicken. It directly interfaces with QuickBooks data, which makes it easy to get your records into the program. For information about TurboTax, visit Intuit's TurboTax Web site at www.turbotax.com or call 800-446-8848.

Kiplinger Tax Cut is produced by Block Financial Corporation of Kansas City, Missouri. Call 800-656-5426 for information.

TaxACT is a product of 2nd Story Software, Marion, Iowa. As this is written, TaxACT does not yet import data from QuickBooks, but may soon. For information visit the company's Web site at www.taxact.com or call 800-573-4287.

Other QuickBooks-related software

Mike Block is a CPA in Florida, who maintains a list of QuickBooks add-on software, on his Internet site (www.blocktax.com). Most of the add-ons in the list are *not* agricultural, but they address a wide variety of business and accounting situations in which QuickBooks is used.

Sources for QuickBooks-Compatible Forms

Here are just a few of the sources for pre-printed paper forms, including checks, invoices, 1099s, W-2s and W-3s, and others:

Intuit:	800-446-8848
Nebs:	800-225-6380
Quill:	800-789-1331
RapidForms:	800-257-8354

Appendix B: QuickBooks List Capacities

The table below shows the maximum capacities of various lists in QuickBooks and QuickBooks Pro (all subject to change in future QuickBooks versions).

List Name	Maximum No. of Entries
Chart of Accounts	10,000
Items	14,500
Customers, Vendors, Employees, and Other Names (combined). *Any of these lists may contain up to 10,000 names, but the combined size of all four lists cannot exceed 14,500 names.*	14,500
Job Types	10,000
Vendor Types	10,000
Customer Types	10,000
Payroll Items	10,000
Classes	10,000
Accounts Receivable Terms and Accounts Payable Terms (combined)	10,000
Payment Methods	10,000
Shipping Methods	10,000
Customer Messages	10,000
To Do	10,000
Memorized Transactions	14,500
Memorized Reports	14,500
Form Templates	14,500

Note: the capacities shown in the table above are the maximum number of list entries *that can ever be created* in each list. In other words, once the maximum number of entries has been created in a list, the list will not allow adding new entries. Deleting old entries from a "full" list *will not* make room for new ones! But as you can see, the numbers must be quite large before there's ever a concern about running out of list space.

Index

C

D

E

F

G

M

ManagePLUS
market value
master note 47
mathematical formulas
memorized transactions
memorizing
menu commands 4
Microsoft Excel
Microsoft Word
milk

N

names
Navigator
non-farm funds
Non-Farm Funds account 243, 245, 247, 249

O

Opening Bal Equity 48
orders

P

P&L
See profit and loss
See Also reports
patronage dividends
Payee Name
payments
payroll
payroll accounts
personal
personal funds
See non-farm funds
petty cash
preferences
principle
printer
printing
production
profit and loss
Profit and Loss report. 323, 325, 327, 329
profit center

CPSIA information can be obtained
at www.ICGtesting.com
Printed in the USA
FFOW04n1709211015
17871FF